THEORY of the EARTH;

or

an INVESTIGATION of the Laws observable in the Composition, Dissolution, and Restoration of Land upon the Globe.

By JAMES HUTTON, M.D. F.R.S. EDIN.

and Member of the Royal Academy of Agriculture at PARIS.

[1788]

Watch BBC
Men of Rock
episode 1
Presented by
Iain Stewart

This book has been published by:

NuVision Publications, LLC
The future of digital publishing.

Contact: sales@nuvisionpublications.com

URL: http://www.nuvisionpublications.com

Publishing Date: 2007

ISBN# 1-59547-777-2

NuVision Publications, LLC

Sioux Falls, SD USA

Contents

THEORY OF THE EARTH: PLATES

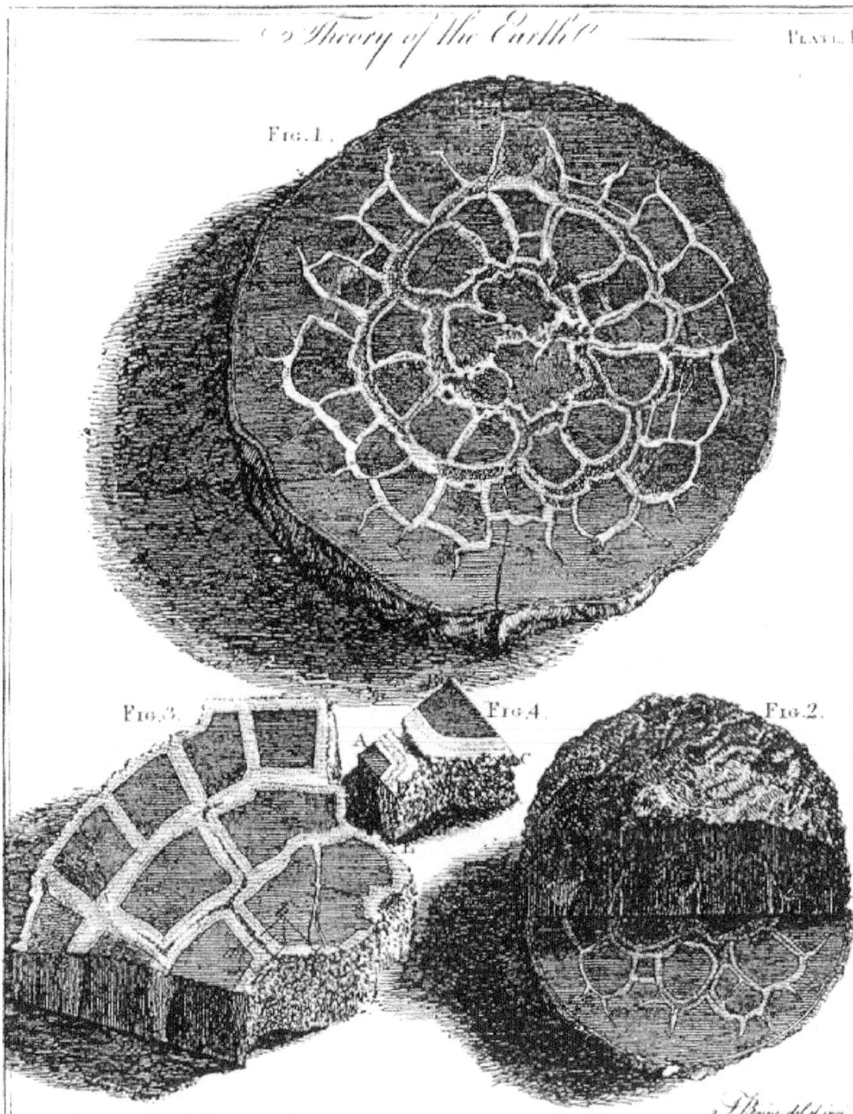

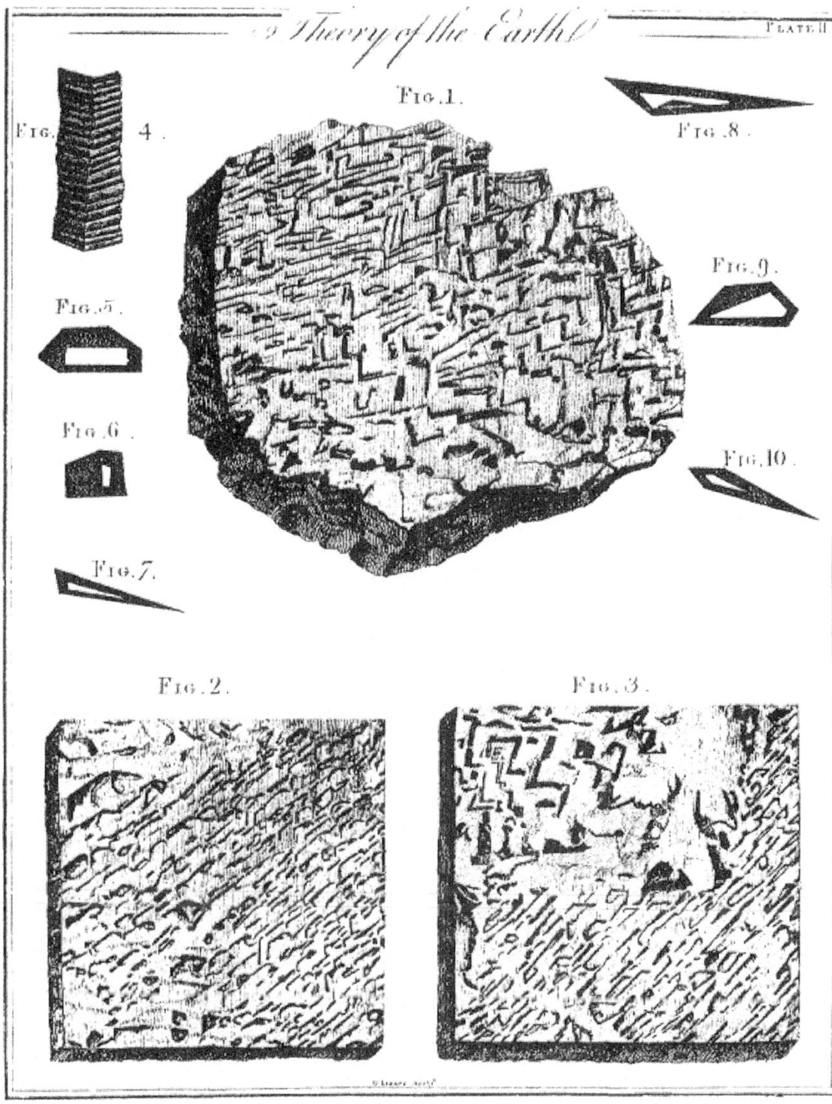
Theory of the Earth
Plate II
Fig. 1.
Fig. 2.
Fig. 3.
Fig. 4.
Fig. 5.
Fig. 6.
Fig. 7.
Fig. 8.
Fig. 9.
Fig. 10.

Praising God

PART. I

Prospect of the Subject to be treated of.

WHEN we trace the parts of which this terrestrial system is composed, and when we view the general connection of those several parts, the whole presents a machine of a peculiar construction by which it is adapted to a certain end. We perceive a fabric, erected in wisdom, to obtain a purpose worthy of the power that is apparent in the production of it.

WE know little of the earth's internal parts, or of the materials which compose it at any considerable depth below the surface. But upon the surface of this globe, the more inert matter is replenished with plants, and with animals and intellectual beings.

WHERE so many living creatures are to ply their respective powers, in pursuing the end for which they were intended, we are not to look for nature in a quiescent state; matter itself must be in motion, and the scenes of life a continued or repeated series of agitations and events.

THIS globe of the earth is a habitable world; and on its fitness for this purpose, our sense of wisdom in its formation must depend. To judge of this point, we must keep in view, not only the end, but the means also by which that end is obtained. These are, the form of the whole, the materials of which it is composed, and the several powers which concur, counter-act, or balance one another, in procuring the general result.

THE form and constitution of the mass are not more evidently calculated for the purpose of this earth as a habitable world, than are the various substances of which that complicated body is composed. Soft and hard parts variously combine, to form a medium consistence adapted to the use of plants and animals; wet and dry are properly mixed for nutrition, or the support of those growing bodies; and hot and cold produce a temperature or climate no less required that a soil. Insomuch, that there is not any particular, respecting either the qualities of the materials, or the construction of the machine, more obvious to our perception, than are the presence and efficacy of design and intelligence in the power that conducts the work.

IN taking this view of things, where ends and means are made the object of attention, we may hope to find a principle upon which the comparative importance of parts in the system of nature may be estimated, and also a rule for selecting the object of our enquiries. Under this direction, science may find a fit subject of investigation in every particular, whether of *form*, *quality*, or *active power*, that presents itself in this system of motion and of life; and which, without a proper attention to this character of the system, might appear anomalous and incomprehensible.

IT is not only by seeing those general operations of the globe which depend upon its peculiar construction as a machine, but also by perceiving how far the particulars, in the construction of that machine, depend upon the general operations of the globe, that we are enabled to understand the constitution of this earth as a thing formed by design. We shall thus also be led to acknowledge an order, not unworthy of Divine wisdom, in a subject which, in another view, has appeared as the work of chance, or as absolute disorder and confusion.

TO acquire a general or comprehensible view of this mechanism of the globe, by which it is adapted to the purpose of forming a habitable world, it is necessary to distinguish three different bodies which compose the whole. These are, a solid body of earth, an aqueous body of sea, and an elastic fluid of air.

IT is the proper shape and disposition of these three bodies that form this globe into a habitable world; and it is the manner in which these constituent bodies are adjusted to each other and the laws of action by which they are maintained in their proper qualities and respective departments, that form the Theory of the machine which we are now to examine.

LET us begin with some general sketch of the particulars now mentioned.

1*st*, THERE is a central body in the globe. This body supports those parts which come to be more immediately exposed to our view, or which may be examined by our sense and observation. This first part is commonly supposed to be solid and inert; but such a conclusion is only mere conjecture; and we shall afterwards find occasion, perhaps, to form another judgment in relation to this subject, after we have examined strictly, upon scientific principles, what appears upon the surface, and have formed conclusions concerning that which must have been transacted in some more central part.

2*dly*, WE find a fluid body of water. This, by gravitation, is reduced to a spherical form, and by the centrifugal force of the earth's rotation, is become oblate. The purpose of this fluid body is essential in the constitution of the world; for, besides affording the means of life and motion to a multifarious race of animals, it is the source of growth and circulation to the organized bodies of this earth, in being the receptacle of the rivers, and the fountain of our vapours.

3*dly*, WE have an irregular body of land, raised above the level of the ocean. This, no doubt, is the smallest portion of the globe; but it is the part to us by far most interesting. It is upon the surface of this part that plants are made to grow; consequently, it is by virtue of this land that animal life, as well as vegetation, is sustained in this world.

Lastly, WE have a surrounding body of atmosphere, which completes the globe. This vital fluid is no less necessary in the constitution of the world than are the other parts; for there is hardly an operation upon the surface of the earth, that is not conducted or promoted by its means. It is a necessary condition for the sustenance of fire; it is the breath of life to animals; it is at least an instrument in vegetation; and while it contributes to give fertility and health to things that grow, it is employed in preventing noxious effects from such as go into corruption. In short, it is the proper means of circulation for the matter of the world by raising up the water of the ocean, and pouring it forth upon the surface of the earth.

SUCH is the mechanism of the globe; let us now mention some of those powers by which motion is produced, and activity procured to the mere machine.

FIRST, there is the progressive force, or moving power, by which this planetary body, if solely actuated, would depart continually from the path which it now pursues, and thus be for ever removed from its end, whether as a planetary body, or as a globe sustaining plants and animals, which may be termed a living world.

BUT this moving body is also actuated by gravitation, which inclines it directly to the central body of the sun. Thus it is made to revolve about that luminary, and to preserve its path.

IT is also upon the same principles, that each particular part upon the surface of the globe, is alternately exposed to the influence of light and darkness, in the diurnal rotation of the earth, as well as in its annual revolution. In this manner are produced the vicissitudes of night and day, so variable in the different latitudes from the equator to the pole, and so beautifully calculated to equalize the benefits of light, so variously distributed in the different regions of the globe.

GRAVITATION and the *vis incita* of matter thus form the first two powers distinguishable in the operations of our system, and wisely adapted to the purpose for which they are employed.

WE next observe the influence of light and heat, of cold and condensation. It is by means of these two powers that the various operations of this living world are more immediately transacted; although the other powers are no less required, in order to produce or modify these great agents in the oeconomy of life, and system of our changing things.

WE do not now enquire into the nature of those powers, or investigate the laws of light and heat, of cold and condensation, by which the various purposes of this world are accomplished; we are only to mention those effects which are made sensible to the common understanding of mankind, and which necessarily imply a power that is employed. Thus, it is by the operation of those powers that the varieties of season in spring and autumn are obtained, that we are blessed with the vicissitudes of summer's heat and winter's cold, and that we possess the benefit of artificial light and culinary fire.

WE are thus bountifully provided with the necessaries of life; we are supplied with things conducive to the growth and preservation of our animal nature, and with fit subjects to employ and to nourish our intellectual powers.

THERE are other actuating powers employed in the operations of this globe, which we are little more than able to enumerate; such are those of electricity and magnetism.

POWERS of such magnitude or force, are not to be supposed useless in a machine contrived surely not without wisdom; but they are mentioned here chiefly on account of their general effect; and it is sufficient to have named powers, of which the actual existence in well known, but of which the proper use in the constitution of the world is still obscure.

WE have thus surveyed the machine in general, with those moving powers, by which its operations, diversified almost *ad infinitum*, are performed. Let us now confine our view, more particularly, to that part of the machine on which we dwell, that so we may consider the natural consequences of those operations which, being within our view, we are better qualified to examine.

THIS subject is important to the human race, to the possessor of this world, to the intelligent being Man, who foresees events to come, and who, in contemplating his future interest, is led to enquire concerning causes, in order that he may judge of events which otherwise he could not know.

IF, in pursuing this object, we employ our skill in research, not in forming vain conjectures; and if *data* are to be found, on which Science may form just conclusions, we should not long remain in ignorance with respect to the natural history of this earth, a subject on which hitherto opinion only, and not evidence, has decided: For in no subject is there naturally less defect of evidence, although philosophers, led by prejudice, or misguided by false theory, have neglected to employ that light by which they should have seen the system of the world.

BUT to proceed in pursuing a little farther our general or preparatory ideas. A solid body of land could not have answered the purpose of a habitable world; for a soil is necessary to the growth of plants; and a soil is nothing but the materials collected from the destruction of the solid land. Therefore, the surface of this land, inhabited by man, and covered with plants and animals, is made by nature to decay, in dissolving from that hard and compact state in which it is found below the soil; and this soil is necessarily washed away, by the continual circulation of the water, running from the summits of the mountains towards the general receptacle of that fluid. THE heights of our land are thus levelled with the shores; our fertile plains are formed from the ruins of the mountains; and those travelling materials are still pursued by the moving water, and propelled along the inclined surface of the earth. These moveable materials, delivered into the sea, cannot, for a long continuance, rest upon the shore; for, by the agitation of the winds, the tides and currents, every moveable thing is carried farther and farther along the shelving bottom of the sea, towards the unfathomable regions of the ocean.

IF the vegetable soil is thus constantly removed from the surface of the land, and if its place is thus to be supplied from the dissolution of the solid earth, as here represented, we may perceive an end to this beautiful machine; an end, arising from no error in its constitution as a world, but from that destructibility of its land which is so necessary in the system of the globe, in the oeconomy of life and vegetation.

THE immense time necessarily required for this total destruction of the land, must not be opposed to that view of future events, which is indicated by the surest facts and most approved principles. Time, which measures every thing in our idea, and is often deficient to our schemes, is to nature endless and as nothing; it cannot limit that by which alone it had existence; and as the natural course of time, which to us seems infinite, cannot be bounded by any operation that may have an end, the progress of things upon this globe, that is, the course of nature, cannot be limited by time, which must proceed in a continual succession. We are, therefore, to consider as inevitable the destruction of our land, so far as effected by those operations which are necessary in the purpose of the globe, considered as a habitable world; and so far as we have not examined any other part of the oeconomy of nature, in which other operations and a different intention might appear.

WE have now considered the globe of this earth as a machine, constructed upon chemical as well as mechanical principles, by which its different parts are all adapted, in form, in quality, and in quantity, to a certain end; an end attained with certainty or success; and an end from which we may perceive wisdom, in contemplating the means employed.

BUT is this world to be considered thus merely as a machine, to last no longer than its parts retain their present position, their proper forms and qualities? Or may it not be also considered as an organized body? Such as has a constitution in which the necessary decay of the machine is naturally repaired, in the exertion of those productive powers by which it had been formed.

THIS is the view in which we are now to examine the globe; to see if there be, in the constitution of this world, a reproductive operation, by which a ruined constitution may be again repaired, and a duration or stability thus procured to the machine, considered as a world sustaining plants and animals.

IF no such reproductive power, or reforming operation, after due enquiry, is to be found in the constitution of this world, we should have reason to conclude, that the system of this earth has either been intentionally made imperfect, or has not been the work of infinite power and wisdom.

HERE is an important question, therefore, with regard to the constitution of this globe; a question which, perhaps, it is in the power of man's sagacity to resolve; and a question which, if satisfactorily resolved, might add some lustre to science and the human intellect.

ANIMATED with this great, this interesting view, let us strictly examine our principles, in order to avoid fallacy in our reasoning; and let us endeavour to support our attention, in developing and subject that is vast in its extent, as well as intricate in the relation of parts to be stated.

THE globe of this earth is evidently made for man. He alone, of all the beings which have life upon this body, enjoys the whole and every part; he alone is capable of knowing the nature of this world, which he thus possesses in virtue of his proper right; and he alone can make the knowledge of this system a source of pleasure and the means of happiness.

MAN alone, of all the animated beings which enjoy the benefits of this earth, employs the knowledge which he there receives, in leading him to judge of the intention of things, as well as of the means by which they are brought about; and he alone is thus made to enjoy, in contemplation as well as sensual pleasure, all the good that may be observed in the constitution of this world; he, therefore, should be made the first subject of enquiry.

NOW, if we are to take the written history of man for the rule by which we should judge of the time when the species first began, that period would be but little removed from the present state of things. The Mosaic history places this beginning of man at no great distance; and there has not been found, in natural history, any document by which a high antiquity might be attributed to the human race. But this is not the case with regard to the inferior species of animals, particularly those which inhabit the ocean and its shores. We find in natural history monuments which prove that those animals had long existed; and we thus procure a measure for the computation of a period of time extremely remote, though far from being precisely ascertained.

IN examining things present, we have data from which to reason with regard to what has been; and, from what has actually been, we have data for concluding with regard to that which is to happen hereafter. Therefore, upon the supposition that the operations of nature are equable and steady, we find, in natural appearances, means for concluding a certain portion of time to have necessarily elapsed, in the production of those events of which we see the effects.

IT is thus that, in finding the relics of sea-animals of every kind in the solid body of our earth, a natural history of those animals is formed, which includes a certain portion of time; and for ascertaining this portion of time, we must again have recourse to the regular operations of the world. We shall thus arrive at facts which indicate a period to which no other species of chronology is able to remount.

IN what follows, therefore, we are to examine the construction of the present earth, in order to understand the natural operations of time past; to acquire principles, by which we may conclude with regard to the future course of things, or judge of those operations, by which a world, so wisely ordered, goes into decay; and to learn, by what means such a decayed world may be renovated, or the waste of habitable land upon the globe repaired.

THIS, therefore, is the object which we are to have in view during this physical investigation; this is the end to which are to be directed all the steps in our cosmological pursuit.

THE solid parts of the globe are, in general, composed of sand, of gravel, of argillaceous and calcareous strata, or of the various compositions of these with some other substances, which it is not necessary now to mention. Sand is separated and sized by streams and currents; gravel is formed by the mutual attrition of stones agitated in water; and marly, or argillaceous strata, have been collected by subsiding in water with which those earthy substances had been floated. Thus, so far as the earth is formed of these materials, that solid body would appear to have been the production of water, winds, and tides.

BUT that which renders the original of our land clear and evident, is the immense quantities of calcareous bodies which had belonged to animals, and the intimate connection of these masses of animal production with the other strata of the land. For it is to be proved, that all these calcareous bodies, from the collection of which the strata were formed, have belonged to the sea, and were produced in it.

WE find the marks of marine animals in the most solid parts of the earth, consequently, those solid parts have been formed after the ocean was inhabited by those animals, which are proper to that fluid medium. If, therefore, we knew the natural history of those solid parts, and could trace the operations of the globe, by which they had been formed, we would have some means for computing the time through which those species of animals have continued to live. But how shall we describe a process which nobody has seen performed, and of which no written history gives any account? This is only to be investigated, *first*, in examining the nature of those solid bodies, the history of which we want to know; and, 2*ndly*, In examining the natural operations of the globe, in order to see if there now actually exist such operations, as, from the nature of the solid bodies, appear to have been necessary to their formation.

BUT, before entering more particularly into those points of discussion, by which the question is to be resolved, let us take a general view of the subject, in order to see what it is which science and observation must decide.

IN all the regions of the globe, immense masses are found, which, though at present in the most solid state, appear to have been formed by the collection of the calcareous *exuviae* of marine animals. The question at present is not, in what manner those collections of calcareous relics have become a perfect solid body, and have been changed from an animal to a mineral substance; for this is a subject that will be afterwards considered; we are now only enquiring, if such is truly the origin of those mineral masses.

THAT all the masses of marble or limestone are composed of the calcareous matter of marine bodies, may be concluded from the following facts:

1*st*, THERE are few beds of marble or limestone, in which may not be found some of those objects which indicate the marine origin of the mass. If, for example, in a mass of marble, taken from a quarry upon the top of the Alps or Andes,[1] there shall be found once cockle-shell , or piece of coral, it must be concluded, that this bed of stone had been originally formed at the bottom of the sea, as much as another bed which is evidently composed almost altogether of cockle-shells and coral. If one bed of limestone is thus found to have been of a marine origin, every concomitant bed of the same kind must be also concluded to have been formed in the same manner.

WE thus shall find the greatest part of the calcareous masses upon this globe to have originated from marine calcareous bodies; for whether we examine marbles, limestones, or such solid masses as are perfectly changed from the state of earth, and are become compact and hard, or whether we examine the soft, earthy, chalky or marly strata, of which so much of this earth is composed, we still find evident proofs, that those beds had their origin from materials deposited at the bottom of the sea; and that they have the calcareous substance which they contain, from the same source as the marbles or the limestones.

2*dly*, IN those calcareous strata, which are evidently of marine origin, there are many parts that are of a sparry structure, that is to say, the original texture of those beds, in such places, has been dissolved, and new structure has been assumed, which is peculiar to a certain state of the calcareous earth. This change is produced by crystallization, in consequence of a previous state of fluidity, which has so disposed the concreting arts, as to allow them to assume a regular shape and structure proper to that substance. A body, whose external form has been modified by this process, is called a *crystal*; one whose internal arrangement of parts is determined by it, is said to be of a *sparry structure*; and this is known from its fracture.

3*dly*, THERE are, in all the regions of the earth, huge masses of calcareous matter, in that crystalline form or sparry state, in which perhaps no vestige can be found of any organized body, nor any

---

[1] "Cette sommité élevée de 984 toises au dessus de notre lac, et par consequent de 1172 au dessus de la mer, est remarquable en ce que l'on y voit des fragmens d'huîtres pétrifiés.---- Cette montagne est dominée par un rocher excarpé, qui s'sil n'est pas inaccessible, est du moins d'un bien difficle accès; il paroît presqu'entierement composé de coquillages pétrifés, renfermés dans un roc calcaire, ou marbre grossier noirâtre. Les fragmens qui s'en detachent, et que l'on rencontre en montant à la Croix de fer, sont remplis de *turbinites* de différentes especes." M. DE SAUSSURE, *Voyage dans les Alpes*, p. 394.

indication that such calcareous matter had belonged to animals; but as, in other masses, this sparry structure, or crystalline state, is evidently assumed by the marine calcareous substances, in operations which are natural to the globe, and which are necessary to the consolidation of the strata, it does not appear, that the sparry masses, in which no figured body is formed, have been originally different from other masses, which, being only crystallized in part, and in part still retaining their original form, leave ample evidence of their marine origin.

WE are led, in this manner, to conclude, that all the strata of the earth, not only those consisting of such calcareous masses, but others superincumbent upon these, have had their origin at the bottom of the sea, by the collection of sand and gravel, of shells, of coralline and crustaceous bodies, and of earths and clays, variously mixed, or separated and accumulated. Here is a general conclusion, well authenticated in the appearance of nature, and highly important in the natural history of the earth.

THE general account of our reasoning is this, that nine tenths, perhaps, or ninety-nine hundredths of this earth, so far as we see, have been formed by natural operations of the globe, in collecting loose materials, and depositing them at the bottom of the sea; consolidating those collections in various degrees, and either elevating those consolidated masses above the level on which they were formed, or lowering the level of the sea.

THERE is a part of the solid earth which we may at present neglect, not, as being persuaded that this part may not also be found to come under the general rule of formation with the rest, but as considering this part to be of no consequence in forming a general rule, which shall comprehend almost the whole, without doing it absolutely. This excluded part consists of certain mountains and masses of granite. These are thought to be still older in their formation, and are very rarely, at least found superincumbent on strata which must be acknowledged as the productions of the sea.

HAVING thus found the greater part, if not the whole, of the solid land to have been originally composed at the bottom of the sea, we may now, in order to form a proper idea of these operations, suppose the whole of this sea-born land to be again dispersed along the bottom of the ocean, the surface of which would rise proportionally over the globe. We would thus have a spheroid of water, with granite rocks and islands scattered here and there. But this would not be the world which we inhabit; therefore, the question now is, how such continents, as we actually have upon the globe, could be erected above the level of the sea.

IT must be evident, that no motion of the sea, caused by this earth revolving in the solar system, could bring about that end; for let us suppose the axis of the earth to be changed from the present poles, and placed in the equinoctial line, the consequence of this might, indeed, be the formation of a continent of land about each new pole, from whence the sea would run towards the new equator; but all the rest of the globe would remain an ocean. Some new points might be discovered, and others, which before appeared above the surface of the sea, would be sunk by the rising of the water; but, on the whole, land could only be gained substantially at the poles. Such a supposition as this, if applied to the present state of things, would be destitute of every support, as being incapable of explaining what appears. BUT even allowing that, by the changed axis of the earth, or any other operation of the globe, as a planetary body revolving in the solar system, great continents of land could have been erected from the place of their formation, the bottom of the sea, and placed in a higher elevation, compared with the surface of that water, yet such a continent as this could not have continued stationary for many thousand years; nor could a continent of this kind have presented to us, every where within its body, masses of consolidated marble and other mineral substances, in a state as different as possible from that in which they were, when originally collected together in the sea.

CONSEQUENTLY, besides an operation, by which the earth at the bottom of the sea should be converted into an elevated land, or placed high above the level of the ocean, there is required, in the operations of the globe, a consolidating power by which the loose materials that had subsided from water should be formed into masses of the most perfect solidity, having neither water nor vacuity between their various constituent parts, nor in the pores of those constituent parts themselves.

HERE is an operation of the globe, whether chemical or mechanical, which is necessarily connected with the formation of our present continents: Therefore, had we a proper understanding of this secret operation, we might thereby be enabled to form an opinion, with regard to the nature of that unknown power, by which the continents have been placed above the surface of that water wherein they had their birth.

IF this consolidating operation be performed at the bottom of the ocean, or under great depths of the earth, of which our continents are composed, we cannot be witnesses to this mineral process, or acquire the knowledge of natural causes, but immediately observing the changes which they produce; but though we have not this immediate observation of those changes of bodies, we have, in science, the means of reasoning from distant events; consequently, of discovering, in the general powers of nature, causes for those events of which we see the effects.

THAT the consolidating operation, in general, lies out of the reach of our immediate observation, will appear from the following truth: All the consolidated masses, of which we now enquire into the cause, are, upon the surface of the earth in a state of general decay, although the various natures of those bodies admit of that dissolution in very different degrees.[2]

FROM every view of the subject, therefore, we are directed to look into those consolidated masses themselves, in order to find principles from whence to judge of those operations by which they had attained their hardness or consolidated state.

IT must be evident, that nothing but the most general acquaintance with the laws of acting substances, and with those of bodies changing by the powers of nature, can enable us to set about this undertaking with any reasonable prospect of success; and here the science of Chemistry must be brought particularly to our aid; for this science, having for its object the changes produced upon the sensible qualities, as they are called, of bodies, by its means we may be enabled to judge of that which is possible according to the laws of nature, and of that which, in like manner, we must consider as impossible.

WHATEVER conclusions, therefore, by means of this science, shall be attained, in just reasoning from natural appearances, this must be held as evidence, where more immediate proof cannot be obtained; and, in a physical subject, where things actual are concerned, and not the imaginations of the human mind, this proof will be considered as amounting to a demonstration.

---

[2] Stalactical and certain ferruginous concretions may seem to form an exception to the generality of this proposition. But an objection of this kind could only arise from a partial view of things; for the concretion here is only temporary, it is in consequence of a solution, and it is to be followed by a dissolution, which will be treated of in its proper place.

PART II.

An Investigation of the Natural Operations employed in consolidating the Strata of the Globe

THERE are just two ways in which porous or spongy bodies can be consolidated, and by which substances may be formed into masses of a natural shape and regular structure; the one of these is simple *congelation* from a fluid state, by means of cold; the other is *accretion*; and this includes a separatory operation, as well as that by which the solid body is to be produced. But, in whichever of these ways solidity is to be procured, it must be brought about by first inducing fluidity, either immediately by the action of heat, or mediately with the assistance of a solvent, that is, but the operation of solution.

THUS, fire and water may be considered as the general agents in this operation which we would explore. We are, therefore, to consider well, what may be the consequences of consolidation by the one or other of those agents; and what may be their several powers with respect to this operation.

IF we are not informed in this branch of science, we may gaze without instruction upon the most convincing proofs of what we want to attain. If our knowledge is imperfect, we may form erroneous principles, and deceive ourselves in reasoning with regard to those works of nature, which are wisely calculated for our instruction.

THE strata, formed at the bottom of the sea, are to be considered as having been consolidated, either by aqueous solution and crystallization, or by the effect of heat and fusion. If it is in the first of these two ways that the solid strata of the globe have attained to their present state, there will be a certain uniformity observable in the effects; and there will be general laws, by which this operation must have been conducted. Therefore, knowing those general laws, and making just observations with regard to the natural appearances of those consolidated masses, a philosopher, in his closet, should be able to determine, what may, and what may not have been transacted in the bowels of the earth , or below the bottom of the ocean.

LET us now endeavour to ascertain what may have been the power of water, acting under fixed circumstances, operating upon known substances, and conducting to a certain end.

THE action of water upon all different substances is an operation with which we are familiar. We have it in our power to apply water in different degrees of heat for the solution of bodies, and under various degrees of compression; consequently, there is no reason to conclude any thing mysterious in the operations of the globe, which are to be performed by means of water, unless an immense compressing power should alter the nature of those operations. But compression alters the relation of evaporation only with regard to heat, or it changes the degree of heat which water may be made to contain; consequently, we are to look for no occult quality in water acting upon bodies at the bottom of the deepest ocean, more than what can be observed in experiments which we have it in our power to try.

WITH regard again to the effect of time. Though the continuance of time may do much in those operations which are extremely slow, where no changes, to our observation, had appeared to take place; yet, where it is not in the nature of things to produce the change in question, the unlimited course of time would be no more effectual, than the moment by which we measure events in our observations.

WATER being the general medium in which bodies collected at the bottom of the sea are always contained, if those masses of collected matter are to be consolidated by solution, it must be by the dissolution of those bodies in that water as a menstruum, and by the concretion or crystallization of this dissolved matter, that the spaces, first occupied by water in those masses, are afterwards to be filled with a hard and solid substance; but without some other power, by which the water contained in those cavities and endless labyrinths of the strata, should be separated in proportion as it had performed its task, it is inconceivable how those masses, however changed from the state of their first subsidence, should be absolutely consolidated, without a particle of fluid water in their composition.

BESIDES this difficulty of having the water separated from the porous masses which are to be consolidated, there is another with which, upon this supposition, we have to struggle. This is, From whence should come the matter with which the numberless cavities in those masses are to be filled?

THE water in the cavities and interstices of those bodies composing strata, must be in a stagnating state; consequently, it can only act upon the surfaces of those cavities which are to be filled up. But with what are they to be filled? Not with water; they are full of this already: Not with the substance of the bodies which contain that water; this would be only to make one cavity in order to fill up another. If, therefore, the cavities of the strata are to be filled with solid matter, but means of water, there must be made to pass through those porous masses, water impregnated with some other substances in a dissolved state; and the aqueous menstruum must be made to separate from the dissolved substance, and to deposit the same in those cavities through which the solution moves.

BY such a supposition as this, we might perhaps explain a partial consolidation of those strata; but this is a supposition, of which the case under consideration does not admit; for in the present case, which is that of materials accumulated at the bottom of the ocean, there is not proper means for separating the dissolved matter from the water included in those enormous masses; nor are there any means by which a circulation in those masses may be formed. In this case, therefore, where the means are not naturally in the supposition, a philosopher, who is to explain the phænomenon by the natural operation of water in this situation, must not have recourse to another agent, still more powerful, to assist his supposition, which cannot be admitted.

THUS, it will appear, that, to consolidate strata formed at the bottom of the sea, in the manner now considered, operations are required unnatural to this place; consequently , not to be supposed in order to support a hypothesis.

BUT now, instead of enquiring how far water may be supposed instrumental in the consolidation of the strata which were originally of loose texture, we are to consider how far there may be appearances in those consolidated bodies, by which it might be concluded, whether or not the present state of their consolidation has been actually brought about by means of that agent.

IF water had been the menstruum by which the consolidating matter was introduced into the interstices of strata, masses of those bodies could only be found consolidated with such substances as water is capable of dissolving; and these substances would be found only in such a state as the simple separation of the dissolving water might produce.

IN this case, the consolidation of strata would be extremely limited; for we cannot allow more power to water than we find it has in nature; nor are we to imagine to ourselves unlimited powers in bodies, on purpose to explain those appearances, by which we should be made to know the powers of nature. Let us, therefore, attend, with every possible circumspection, to the appearances of those bodies, by means of which we are to investigate the principles of mineralogy, and know the laws of nature.

THE question now before us concerns the consolidating substances of strata. Are these such as will correspond to the dissolving power of water, and to the state in which those substances might be left by the separation of their menstruum? No; far, far from this supposition is the conclusion that necessarily follows from natural appearances.

WE have strata consolidated by calcareous spar, a thing perfectly distinguishable from the stalactical concretion of calcareous earth, in consequence of aqueous solution. We have strata made solid by the formation of fluor, a substance not soluble, so far as we know, by water. We have strata consolidated with sulphureous and bituminous substances, which do not correspond to the solution of water. We have strata consolidated with siliceous matter, in a state totally different from that under which it has been observed, on certain occasions, to be deposited by water. We have strata consolidated by feldspar, a substance insoluble in water. We have strata consolidated by almost all the various metallic substances, with their almost endless mixtures and sulphureous compositions; that is to say, we find, perhaps, every different substance introduced into the interstices of strata which had been formed by subsidence at the bottom of the sea.

IF it is by means of water that those interstices have been filled with those materials, water must be, like fire, an universal solvent, or cause of fluidity, and we must change entirely our opinion of water in relation to its chemical character. But there is no necessity thus to violate our chemical principles, in order to explain certain natural appearances; more especially if those appearances may be explained in another manner, consistently with the known laws of nature.

IF, again, it is by means of heat and fusion that the loose and porous structure of strata shall be supposed to have been consolidated, then every difficulty which had occurred in reasoning upon the power or agency of water is at once removed. The loose and discontinuous body of a stratum may be closed by means of softness and compression; the porous structure of the materials may be consolidated, in a similar manner, by the fusion of their substance; and foreign matter may be introduced into the open structure of strata, in form of steam or exhalation, as well as in the fluid state of fusion; consequently, heat is an agent competent for the consolidation of strata, which water alone is not. If, therefore, such an agent could be found acting in the natural place of strata, we must pronounce it proper to bring about that end.

THE examination of nature gives countenance to this supposition, so far as strata are found consolidated by every species of substance, and almost every possible mixture of those different substances; consequently, however difficulty it may appear to have this application of heat, for the purpose of consolidating strata formed at the bottom of the ocean, we cannot, from natural appearances, suppose any other cause, as having actually produced the effects which are now examined.

THIS question, with regard to the means of consolidating the strata of the globe, is, to natural history, of the greatest importance; and it is essential in the theory now proposed to be given of the mineral system. It would, therefore, require to be discussed with some degree of precision, in examining the particulars; but of these, there is so great a field, and the subject is so complicated in its nature, that volumes might be written upon particular branches only, without exhausting what might be said upon the subject; because the evidence, though strong in many particulars, is chiefly to be enforced by a multitude of facts, conspiring, in a diversity of ways, to point out one truth, and by the impossibility of reconciling all these facts, except by means of one supposition.

BUT, as it is necessary to give some proof of that which is to be a principle in our reasoning afterwards, I shall now endeavour to generalize the subject as much as possible, in order to answer that end, and, at the same time, to point out the particular method of enquiry. THERE are to be found, among the various strata of the globe, bodies formed of two different kinds of substances, *siliceous* bodies, and those which may be termed *sulphureous.* With one or other, or both of those two substances, every different consolidated stratum of the globe will be found so intimately mixed, or closely connected, that it must be concluded, by whatever cause those bodies of siliceous and sulphureous matter had been changed from a fluid to a concreted state, the strata must have been similarly affected by the same cause.

THESE two species of bodies, therefore, the siliceous and the sulphureous, may now be examined, in relation to the causes of their concretion, with a view to determine, what has been the general concreting or consolidating power, which has operated universally in the globe; and particularly to shew, it has not been by means of any fluid solution, that strata in general have been consolidated, or that those particular substances have been crystallized and concreted.

SILICEOUS matter, physically speaking, is not soluble in water; that is to say, in no manner of way have we been enabled to learn, that water has the power of dissolving this matter.

MANY other substances, which are so little soluble in water, that their solubility could not be otherwise detected of themselves, are made to appear soluble by means of siliceous matter; such is fel-spar, one of the component parts of rock-granite.

FELD-SPAR is a compound of siliceous, argillaceous, and calcareous earth, intimately united together. This compound siliceous body being, for ages, exposed to the weather, the calcareous part of it is dissolved, and the siliceous part is left in form of a soft white earth. But whether this dissolution is performed by pure water, or by means also of an acid, may perhaps be questioned. This, however, is certain, that we must consider siliceous substances as insoluble in water. THE water of Giezer in Iceland undoubtedly contains this substance in solution; but there is no reason to believe, that it is here dissolved by any other than natural means; that is, an alkaline substance, by which siliceous bodies may be rendered soluble in water.

IT may be, therefore, asserted, that no siliceous body having the hardness of flint, nor any crystallization of that substance, has ever been formed, except by fusion. If, by any art, this substance shall be dissolved in simple water, or made to crystallize from any solution, in that case, the assertion which has been here made may be denied. But where there is not the vestige of any proof, to authorise the supposition of flinty matter being dissolved by water, or crystallized from that solution, such an hypothesis cannot be admitted, in opposition to general and evident appearances.

BESIDES this proof for the fusion of siliceous bodies, which is indirect, arising from the indissolubility of that substance in water, there is another, which is more direct, being founded upon appearances which are plainly inconsistent with any other supposition, except that of simple fluidity induced by heat. The proof I mean is, the penetration of many bodies with a flinty substance, which, according to every collateral circumstance, must have been performed by the flinty matter in a simply fluid state, and not in a state of dissolution by a solvent.

THESE are flinty bodies perfectly insulated in strata both of chalk and sand. It requires but inspection to be convinced. It is not possible that flinty matter could be conveyed into the middle of those strata, by a menstruum in which it was dissolved, and thus deposited in that place, without the smallest trace of deposition in the surrounding parts.

BUT, besides this argument taken from what does not appear, the actual form in which those flinty masses are found, demonstrates, *first*, That they have been introduced among those strata in a fluid state, by injection from some other place. 2*dly*, That they have been dispersed in a variety of ways among those strata, then deeply immersed at the bottom of the sea; and, *lastly*, That they have been there congealed from the state of fusion, and have remained in that situation, while those strata have been removed from the bottom of the ocean to the surface of the present land.

TO describe those particular appearances would draw this paper beyond the bounds of an essay. We must, therefore, refer those who would enquire more minutely into the subject, to examine the chalk-countries of France and England, in which the flint is found variously formed; the sand-hills interspersed among those chalk-countries, which have been also injected by melted flint; and the pudding-stone of England, which I have not seen in its natural situation. More particularly, I would recommend an examination of the insulated masses of stone, found in the sand-hills by the city of Brussels; a stone which is formed by an injection of flint among sand, similar to that which, in a body of gravel, had formed the pudding-stone of England .[3]

ALL these examples would require to be examined upon the spot, as a great part of the proof for the fusion of the flinty substances, arises, in my opinion, from the form in which those bodies are found, and the state of the surrounding parts. But there are specimens brought from many different places, which contain, in themselves, the most evident marks of this injection of the flinty substance in a fluid state. These are pieces of fossil wood, penetrated with a siliceous substance, which are brought from England, Germany, and Lochneagh in Ireland.

IT appears from these specimens, that there has sometimes been a prior penetration of the body of wood, either with irony matter, or calcareous substance. Sometimes, again, which is the case with that of Lochneagh, there does not seem to have been any penetration of those two substances. The injected flint appears to have penetrated the body of this wood, immersed at the bottom of the sea, under an immense compression of water. This appears from the wood being penetrated partially, some parts not being penetrated at all.

---

[3] ACCURATE descriptions of those appearances, with drawings, would be, to natural history, a valuable acquisition.

NOW, in the limits between those two parts, we have the most convincing proof, that it had been flint in a simple fluid state which had penetrated the wood, and not in a state of solution.

First, BECAUSE, however little of the wood is left unpenetrated, the division is always distinct between the injected part of that which is not penetrated by the fluid flint. In this case, the flinty matter has proceeded a certain length, which is marked, and no farther; and, beyond this boundary, there is no partial impregnation, nor a gradation of the flintifying operation, as must have been the case if siliceous matter had been deposited from a solution. *2dly*, The termination of the flinty impregnation has assumed such a form, precisely, as would naturally happen from a fluid flint penetrating that body.

IN other specimens of this mineralizing operation, fossil wood, penetrated, more of less, with ferruginous and calcareous substances, has been afterwards penetrated with a flinty substance. In this case, with whatever different substances the woody body shall be supposed to have been penetrated in a state of solution by water, the regular structure of the plant would still have remained, with its vacuities variously filled with the petrifying substances, separated from the aqueous menstruum, and deposited in the vascular structure of the wood.

THERE cannot be a doubt with regard to the truth of this proposition; for as it is, we frequently find parts of the consolidated wood, with the vascular structure remaining perfectly in the natural shape and situation; but if it had been by aqueous solution that the wood had been penetrated and consolidated, all the parts of that body would be found in the same natural shape and situation.

THIS, however, is far from being the case; for while, in some parts, the vascular structure is preserved entire, it is also evident, that, in general, the woody structure is variously broken and dissolved by the fusion and crystallization of the flint. There are so many and such various convincing examples of this, that, to attempt to describe them, would be to exceed the bounds prescribed for this dissertation; but such specimens are in my possession, ready for the inspection of any person who may desire to study the subject.

WE may now proceed to consider sulphureous substances, with regard to their solubility in water, and to the part which these bodies have acted in consolidating the strata of the globe.

THE sulphureous substances here meant to be considered, are substances not soluble in water, so far as we know, but fusible by heat, and inflammable by means of heat and vital air. These substances are of two kinds; the one more simple, the other more compound.

THE most simple kind is composed of two different substances, *viz.* phlogiston, with acid or metallic substances; from which result, on the one hand, sulphur, and, on the other, metals, both properly so called.

The more compound sort, again, is oily matter, produced by vegetables, and forming bituminous bodies.

THE *first* of these is found naturally combined with almost all metallic substances, which are then said to be mineralized with sulphur. Now, it is well known, that this mineralizing operation is performed by means of heat or fusion; and there is no person skilled in chemistry that will pretend to say, this may be done by aqueous solution. The combination of iron and sulphur, for example, may easily be performed by fusion; but, by aqueous solution, this particular combination is again resolved, and forms an acido-metallic, that is, a vitriolic substance, after the phlogiston (which refuses aqueous solution) has been separated from the composition, by means of the joint operation of vital air.

THE variety of these sulphureo-metallic substances, in point of composition, is almost indefinite; but, unless they were all soluble in water, this could not have happened by the action of that solvent. If we shall allow any one of those bodies to have been formed by the fluidity of heat, they must all have been formed in the same manner; for there is such a chain of connection among those bodies in the mineral regions, that they must all have been composed, either, on the one hand, by aqueous solution, or, on the other, by means of heat and fusion.

HERE, for example, are crystallized together in one mass, *first, Pyrites*, containing sulphur, iron copper; 2*dly, Blend*, a composition of iron, sulphur, and calamine; 3*rdly, Galena*, consisting of lead and sulphur; 4*thly, Marmor metallicum*, being the terra ponderosa, saturated with the vitriolic acid; a substance insoluble in water; 5*thly, Fluor*, a saturation of calcareous earth, with a peculiar acid, called the acid of spar, also insoluble in water; 6*thly, Calcareous spar*, of different kinds, being calcareous earth saturated with fixed air, and something besides, which forms a variety in this substance; *lastly, Siliceous substances*, or *Quartz crystals*. All these bodies, each possessing its proper shape, are mixed in such a manner as it would be endless to describe, but which may be expressed in general by saying, that they are mutually contained in, and contain each other.

UNLESS, therefore, every one of these different substances may be dissolved in water, and crystallized from it, it is in vain to look for the explanation of these appearances in the operations of nature, by the means of aqueous solution. ON the other hand, heat being capable of rendering all these substances liquid, they may be, with the greatest simplicity, transported from one place to another; and they may be made to concrete altogether, at the same time, and distinctly separate in any place. Hence, for the explanation of those natural appearances, which are so general, no further conditions are required, than the supposition of a sufficient intensity of subterraneous fire or heat, and a sufficient degree of compression upon those bodies, which are to be subjected to that violent heat, without calcination or change. But, so far as this supposition is not gratuitous, the appearances of nature will be thus explained.

I SHALL only mention one specimen, which must appear most decisive of the question. It is, I believe, from an Hungarian mine. In this specimen, petro-silex, pyrites, and cinnabar, are so mixed together, and crystallized upon each other, that it is impossible to conceive any one of those bodies to have had its fluidity and concretion from a cause which had not affected the other two. Now, let those who would deny the fusion of this siliceous body explain how water could dissolved these three different bodies, and deposit them in their present shape. If, on the contrary, they have not the least shadow of reason for such gratuitous supposition, the present argument must be admitted in its full force.

SULPHUR and metals are commonly found combined in the mineral regions. But this rule is not universal; for they are also frequently in a separate state. There is not, perhaps, a metal, among the great number which are now discovered, that may not be found native, as they are called, or in their metallic state.

METALLIC substances are also thus found in some proportion to the disposition of the particular metals, to resist the mineralizing operations, and to their facility of being metallized by fire and fusion. Gold, which refuses to be mineralized with sulphur, is found generally in its native state. Iron, again, which is so easily mineralized and scorified, is seldom found its malleable state. The other metals are all found more of less mineralized, though some of them but rarely in the native state.

BESIDES being found with circumstances thus corresponding to the natural facility, or to the impediments attending the metallization of those different calces, the native metals are also found in such a shape, and with such marks, as can only agree with the fusion of those bodies; that is to say, those appearances are perfectly irreconcileable with any manner of solution and precipitation.

FOR the truth of this assertion, among a thousand other examples, I appeal to that famous mass of native iron, discovered by Mr PALLAS, in Siberia. This mass being so well known to all the mineralists of Europe, any comment upon its shape and structure will be unnecessary.[4] WE

---

[4] Since this Dissertation was written, M. DE LA PEYROUSE has discovered a native manganese. The circumstances of this mineral are so well adapted for illustrating the present doctrine, and so well related by M. DE LA PEYROUSE, that I should be wanting to the interest of mineral knowledge, were I not to give here that part of his Memoir.

"LORSQUE je fis insérer dans the journal de physique de l'anneée 1780, au mois de Janvier, une Dissertation contenant la classification des mines de manganèse, je ne connoissois point, à cette epoque, la mine de manganèse native. Elle a la couleur de son régule: elle salit les doigts de la même teinte. Son tissu paroit aussi lamelleux, et les lames semblent affecter une sorte de divergence. Elle a ainsi que lui, l'éclat métalllique; comme lui elle se laisse applatir sous le marteau, et s'exfolie si l'on redouble les coups; mais une circonstance qui est trop frappante pour que je l'omette, c'est la figure de la manganèse native, si prodigieusement conforme à celle du régule, qu'on s'y laisseroit tromper, si la mine n'étoit encore dans sa gangue: figure très-essentielle à observer ici, parce qu'elle est due à la nature même de la manganèse. En effect, pour réduire toutes les mines en général, il faut employer divers flux

come now to the *second* species of inflammable bodies called oily or bituminous. These substances are also found variously mixed with mineral bodies, as well as forming strata of themselves; they are, therefore, a proper subject for a particular examination.

IN the process of vegetation, there are produced oily and resinous substances; and from the collection of these substances at the bottom of the ocean, there are formed strata, which have been variously changed, in consequence of the effects of that heat, according as the distillation of the more volatile parts of those bodies has been suffered to proceed.

IN order to understand this, it must be considered that, while immersed in water, and under insuperable compression, the vegetable, oily, and resinous substances, would appear to be unalterable by heat; and it is only in proportion as certain chemical separations take place, that these inflammable bodies are changed in their substance by the application of heat. Now, the most general change of this kind is in consequence of evaporation, or the distillation of their more volatile parts, by which oily substances become bituminous, and bituminous substances become coaly.

THERE is here a gradation which may best understood by comparing the extremes. ON the one hand, we know by experiment, that oily and bituminous substances can be melted and partly changed into vapour by heat, and that they become harder and denser, in proportion as the more volatile parts have evaporated from them. On the other hand, coaly substances are destitute of fusibility and volatility, in proportion as they have been exposed to greater degrees of heat, and to other circumstances favourable to the dissipation of their more volatile and fluid parts.

IF, therefore, in mineral bodies, we find the two extreme states of this combustible substance, and also the intermediate states, we must either conclude, that this particular operation of heat has been thus actually

---

appropriés. Pour la reduction de la manganèse, bien loin d'user de ce moyen, il faut, au contraire, éloigner tout flux, produire la fusion, par la seule violence et la promptetude du feu. Et telle est la propension naturelle et prodigieuse de la manganèse à la vitrification, qu'on n'a pu parvenir encore à réduire son régule en un seul culot; on trouve dans le creuset plusieurs petits boutons, qui forment autant de culots p. 239 séparés. Dans la mine de manganèse native, elle n'est point en une seule masse; elle est disposée également en plusieurs culots séparés, et un peu applatis, comme ceux que l'art produit; beaucoup plus gros, à la vérité, parce que les agens de la nature doivent avoir une autre énergie, que ceux de nos laboratoires; et cette ressemblance si exacte, semble devoir vous faire penser que la mine native a été produite par le feu, tout comme son régule. La presence de la chaux argentée de la manganèse, me permettroit de croire que la nature n'a fait que réduire cette chaux. Du reste, cette mine native est très-pure, et ne contient aucune partie attirable à l'aimant. Cette mine, unique jusqu'à ce moment, vient, tout comme les autres manganèse que j'ai décrites, des mines de fer de *Sem*, dans la vallée de *Viedersos*, en Comté de Foix." *Journal de Physique, Janvier 1786.*

employed in nature, or we must explain those appearances by some other means, in as satisfactory a manner, and so as shall be consistent with other appearances.

IN this case, it will avail nothing to have recourse to the false analogy of water dissolving and crystallizing salts, which has been so much employed for the explanation of other mineral appearances. The operation here in question is of a different nature, and necessarily requires both the powers of heat and proper conditions for evaporation.

THEREFORE, in order to decide the point, with regard to what is the power in nature by which mineral bodies have become solid, we have but to find bituminous substance in the most complete state of coal, intimately connected with some other substance, which is more generally found consolidating the strata, and assisting in the concretion of mineral substances. But I have in my possession the most undoubted proof of this kind. It is a mineral vein, or cavity, in which are blended together coal of the most fixed kind, quartz and marmor metallicum. Nor is this all; for the specimen now referred to is contained in a rock of this kind, which every naturalist now-a-days will allow to have congealed from a fluid state of fusion. I have also similar specimens from the same place, in which the coal is not of that fixed and infusible kind, which burns without flame or smoke, but is bituminous or inflammable coal.

WE have hitherto been resting the argument upon a single point, for the sake of simplicity or clearness, not for want of those circumstances which shall be found to corroborate the theory. The strata of fossil coal are found in almost every intermediate state, as well as in those of bitumen and charcoal. Of the one kind is that fossil coal which melts or becomes fluid upon receiving heat; of the other, is that species of coal, found both in Wales and Scotland, which is perfectly infusible in the fire, and burns like coaks, without flame or smoak. The one species abounds in oily matter, the other has been distilled by heat, until it has become a *caput mortuum*, or perfect coal.

THE more volatile parts of these bituminous bodies are found in their separate state on some occasions. There is a stratum of limestone in Fifeshire near Raith, which, though but slightly tinged with a black colour, contains bituminous matter, like pitch, in many cavities, which are lined with calcareous spar crystallized. I have a specimen of such a cavity, in which the bitumen is in sphericles, or rounded drops, immersed in the calcareous spar.

NOW, it is to be observed, that, if the cavity in the solid limestone or marble, which is lined with calcareous crystals containing pyrites, had been thus encrusted by means of the filtration of water, this water must have dissolved calcareous spar, pyrites and bitumen. But these natural appearances would not even be explained by this dissolution and supposed filtration of those substances. There is also required, *first*, a cause for the separation of those different substances, form the aqueous menstruum in which they had been dissolved: *2dly*, An explanation of the way in which a dissolved bitumen should be formed into round hard bodies of the most solid structure; and, *lastly*, Some probable means for this complicated operation being performed, below the bottom of the ocean, in the close cavity of a marble stratum.

THUS, the additional proof, from the facts relating to the bituminous substances, conspiring with that from the phaenomena of other bodies, affords the strongest corroboration of this opinion, that the various concretions found in the internal parts of strata have not been occasioned by means of aqueous solution, but by the power of heat and operation of simple fusion, preparing those different substances to concrete and crystallize in cooling.

THE arguments which have been now employed for proving that strata have been consolidated by the power of heat, or by the means of fusion, have been drawn chiefly from the insoluble nature of those consolidating substances in relation to water, which is the only general menstruum that can be allowed for the mineral regions. But there are found in the mineral kingdom, many solid masses of sal gem, which is a soluble substance. It may be now enquired, how far these masses, which are not unfrequent in the earth, tend either to confirm the present theory, or, on the contrary, to give countenance to that which supposes water the chief instrument in consolidating strata.

THE formation of salt at the bottom of the sea, without the assistance of subterranean fire, is not a thing unsupposable, as at first sight it might be. Let us but suppose a rock place across the gut of Gibraltar, (a case nowise unnatural), and the bottom of the Mediterranean would be certainly filled with salt, because the evaporation from the surface of that sea exceeds the measure of its supply.

BUT strata of salt, formed in this manner at the bottom of the sea, are as far from being consolidated by means of aqueous solution, as a bed of sand in the same situation; and we cannot explain the consolidation of such a stratum of salt by means of water, without supposing subterranean heat employed, to evaporate the brine which would successively occupy the interstices of the saline crystals. But this, it may be observed, is equally departing from the natural operation of water, as the means for consolidating the sediment of the ocean, as if we were to suppose the same thing done by heat and fusion. For the question is not, if subterranean heat be of sufficient intensity for the purpose of consolidating strata by the fusion of their substances; the question, is whether it be by means of this agent, subterranean heat, or by water alone, without the operation of a melting heat, that those materials have been variously consolidated.

THE example now under consideration, consolidated mineral salt, will serve to throw some light upon the subject; for as it is to be shewn, that this body of salt had been consolidated by perfect fusion, and not by means of aqueous solution, the consolidation of strata of dissoluble substances, by the operation of a melting heat, will meet with all that confirmation which the consistency of natural appearances can give.

THE rock salt in Cheshire lies in strata of red marl. It is horizontal in its direction. I do not know its thickness, but it is dug thirty or forty feet deep. The body of this rock is perfectly solid, and the salt, in many places, pure, colourless and transparent, breaking with a sparry cubical structure. But the greatest part is tinged by the admixture of the marl, and that in various degrees, from the slightest tinge of red, to the most perfect opacity. Thus, the rock appears as if it had been a mass of fluid salt, in which had been floating a quantity of marly substance, not uniformly mixed, but every where separating and subsiding from the pure saline substance.

THERE is also to be observed a certain regularity in this separation of the tinging from the colourless substance, which, at a proper distance, gives to the perpendicular section of the rock a distinguishable figure in its structure. When looking at this appearance near the bottom of the rock, it, at first, presented me with the figure of regular stratification; but, upon examining the whole mass of rock, I found, that it was only towards the bottom that this stratified appearance took place; and that, at the top of the rock, the most beautiful and regular figure was to be observed; but a figure the most opposite to that of stratification. It was all composed of concentric circles; and these appeared to be the section of a mass, composed altogether of concentric spheres, like those beautiful systems of configuration which agates so frequently present us with in miniature. In about eight or ten feet from the top, the circles growing large, were blended together, and gradually lost their regular appearance, until, at a greater depth, they again appeared in resemblance of a stratification.

THIS regular arrangement of the floating marly substance in the body of salt, which is that of the structure of a coated pebble, or that of concentric spheres, is altogether inexplicable upon any other supposition, than the perfect fluidity or fusion of the salt, and the attractions and repulsions of the contained substances. It is in vain to look, in the operations of solution and evaporation, for that which nothing but perfect fluidity or fusion can explain.

THIS example of a mineral salt congealed from a melted state, may be confirmed from another which I have from Dr BLACK, who suggested it to me. It is an alkaline salt, found in a mineral state, and described in the Philosophical Transactions, *anno* 1771. But to understand this specimen, something must be premised with regard to the nature of fossil alkali.

THE fossil alkali crystallizes from a dissolved state, in combining itself with a large portion of the water, in the manner of alum; and, in this case, the water is essential to the constitution of that transparent crystalline body; for, upon the evaporation of the water, the transparent salt loses its solidity, and becomes a white powder. If, instead of being gently dried, the crystalline salt is suddenly exposed to a sufficient degree of heat, that is, somewhat more than boiling water, it enters into the state of aqueous fusion, and it boils, in emitting the water by means of which it had been crystallized in the cold, and rendered fluid in that heated state. It is not possible to crystallize this alkaline salt from a dissolved state, without the combination of that quantity of water, nor to separate that water without destroying its crystalline state.

BUT in this mineral specimen, we have a solid crystalline salt, with a structure which, upon fracture, appears to be sparry and radiated, something resembling that of zeolite. It contains no water in its crystallization, but melts in a sufficient heat, without any aqueous fusion. Therefore, this salt must have been in a fluid state of fusion, immediately before its congelation and crystallization.

IT would be endless to give examples of particular facts, so many are the different natural appearances that occur, attended with a variety of different circumstances.

THERE is one, however, which is peculiarly distinct, admits of sufficiently accurate description, and contains circumstances from which conclusions may be drawn with clearness. This is the iron-stone, which is commonly found among the argillaceous strata, attendant upon fossil coal, both in Scotland and in England.

THIS stone is generally found among the bituminous schistus, or black argillaceous strata, either in separate masses of various shapes and sizes, or forming of itself strata with are more or less continuous in their direction among the schistus or argillaceous beds.

THIS mineral contains in general from 40 to 50 *per cent.* of iron, and it loses near one third of its weight in calcination. Before calcination it is of a gray colour, is not penetrable by water, and takes a polish. In this state, therefore, it is perfectly solid; but being calcined, it becomes red, porous, and tender. THE fact to be proved with regard to these iron-stones is this, That they have acquired their solid state from fusion, and not in concreting from any aqueous solution.

TO abridge this disquisition, no argument is to be taken from contingent circumstances, (which, however, are often found here as well as in the case of marbles); such only are to be employed as are general to the subject, and arise necessarily from the nature of the operation.

IT will be proper to describe a species of these stones, which is remarkably regular in its form. It is that found at Aberlady in East Lothian.

THE form of these iron-stones is that of an oblate or much compressed sphere, and the size from two or three inches diameter to more than a foot. In the circular or horizontal section, they present the most elegant septarium;[5] and, from the examination of this particular structure, the following conclusions may be drawn.

---

[5] Plate I.

First, THAT the septa have been formed by the uniform contraction of the internal parts of the stone, the volume of the central parts diminishing more than that of the circumference; by this means, the separations of the stone diminish, in a progression from the centre towards the circumference.

2*d*, THAT there are only two ways in which the septa must have received the spar with which they are filled, more or less, either, *first*, By insinuation into the cavity of the septa after these were formed; or, 2*dly*, By separation from the substance of the stone, at the same time that the septa were forming.

WERE the first supposition true, appearances would be observable, shewing that the sparry substance had been admitted, either through the porous structure of the stone, or through proper apertures communicating from without. Now, if either one or other of these had been the case, and that the stone had been consolidated from no other cause than concretion from a dissolved state, that particular structure of the stone, by means of which the spar had been admitted, must appear at present upon an accurate examination.

THIS, however, is not the case, and we may rest the argument here. The septa reach not the circumference; the surface of the stone is solid and uniform in every part; and there is not any appearance of the spar in the argillaceous bed around the stone.

IT, therefore, necessarily follows, that the contraction of the iron-stone, in order to form septa, and the filling of these cavities with spar, had proceeded *pari passu*; and that this operation must have been brought about by means of fusion, or by congelation from a state of simple fluidity and expansion.

IT is only further to be observed, that all the arguments which have been already employed, concerning mineral concretions from a simply fluid state, or that of fusion, here take place. I have septaria of this kind, in which, besides pyrites, iron-ore, calcareous spar, and another that is ferruginous and compound, there is contained siliceous crystals; a case which is not so common. I have them also attended with circumstances of concretion and crystallization, which, besides being extremely rare, are equally curious and interesting.

THERE is one fact more which is well worth our attention, being one of those which are so general in the mineral regions. It is the crystallizations which are found in the close cavities of the most solid bodies.

NOTHING is more common than this appearance. Cavities are everywhere found closely lined with crystallization, of every different substance which may be supposed in those places. These concretions are well known to naturalists, and form part of the beautiful specimens which are preserved in the cabinets of collectors, and which the German mineralists have termed *Drusen*. I shall only particularize one species, which may be described upon principle, and therefore may be a proper subject on which to reason, for ascertaining the order of production in certain bodies. This body, which we are now to examine, is of the agate species.

WE have now been considering the means employed by nature in consolidating strata which were originally of an open structure; but in perfectly solid strata, we find bodies of agate, which have evidently been formed in that place where they now are found. This fact, however, is not still that of which we are now particularly to enquire; for this, of which we are to treat, concerns only a cavity within this agate; now, whatever may have been the origin of the agate itself, we are to shew, from what appears within its cavity, that the crystallizations which are found in this place had arisen from a simply fluid state, and not from that of any manner of solution.

THE agates now in question are those of the coated kind, so frequent in this country, called pebbles. Many of these are filled with a siliceous crystallization, which evidently proceeds from the circumferences towards the centre. Many of them, again, are hollow. Those cavities are variously lined with crystallized substances; and these are the object of the present examination.

BUT before describing what is found within, it is necessary to attend to this particular circumstance, that the cavity is perfectly inclosed with many solid coats, impervious to air or water, but particularly with the external cortical part, which is extremely hard, takes the highest polish, and is of the most perfect solidity, admitting the passage of nothing but light and heat.

WITHIN these cavities, we find, *first*, The coat of crystals with which this cavity is always lines; and this is general to all substances concreting, in similar circumstances, from a state of fusion; for when thus at liberty they naturally crystallize. *2dly*, We have frequently a subsequent crystallization, set upon the first, and more or less immersed in it. *3dly*, There is also sometimes a third crystallization, superincumbent on the second, and in like manner as the second is on the first. I shall mention some particulars.

I HAVE one specimen, in which the primary crystals are siliceous, the secondary thin foliaceous crystals of deep red but transparent iron-ore, forming elegant figures, that have the form of roses. The tertiary crystallization is a frosting of small siliceous crystals upon the edges of the foliaceous crystals.

IN other specimens, there is first a lining of colourless siliceous crystals, then another lining of amethystine crystals, and sometimes within that, fuliginous crystals. Upon these fuliginous and amethystine crystals are many sphericles or hemispheres of red compact iron-ore, like haematites.

IN others, again, the primary crystals are siliceous, and the secondary calcareous. Of this kind, I have one which has, upon the calcareous crystals, beautiful transparent siliceous crystals, and iron-sphericles upon these.

Lastly, I HAVE an agate formed of various red and white coats, and beautifully figured. The cavity within the coated part of the pebble is filled up without vacuity, first, with colourless siliceous crystals; secondly, with fuliginous crystals; and, lastly, with white or colourless calcareous spar. But between the spar and crystals there are many sphericles, seemingly of iron, half sunk into each of these two different substances.

FROM these facts, I may now be allowed to draw the following conclusions:

First, THAT concretion had proceeded from the surface of the agate body inwards. This necessarily follows from the nature of those figured bodies, the figures of the external coats always determining the shape of those within, and never, contrarily, those within affecting those without.

2dly, THAT when the agate was formed, the cavity then contained everything which now is found within it, and nothing more.

3dly, THAT the contained substances must have been in a fluid state, in order to their crystallizing.

Lastly, THAT as this fluid state had not been the effect of solution in a menstruum, it must have been fluidity from heat and fusion.

THERE are in jaspers and agates many other appearances, from whence this last conclusion may be formed with great certainty and precision; but it is hoped, that what has been now given may suffice for establishing that proposition without any doubt.

IT must not here be objected, that there are frequently found siliceous crystals and amethysts containing water; and that it is impossible to confine water even in melted glass. It is true, that here, at the surface of the earth, melted glass cannot, in ordinary circumstances, be made to receive and inclose condensed water; but let us only suppose a sufficient degree of compression in the body of melted glass, and we can easily imagine it to receive and confine water, as well as any other substance. But if, even in our operations, water, by means of compression, may be made to endure the heat of red hot iron without being converted into vapour, what may not the power of nature be able to perform? The place of mineral operations is not on the surface of the earth; and we are not to limit nature with our imbecility, or estimate the powers of nature by the measure of our own.

TO conclude this long chemico-mineral disquisition, I have specimens in which the mixture of calcareous, siliceous and metallic substances, in almost every species of concretion which is to be found in mineral bodies, may be observed, and in which there is exhibited, in miniature, almost every species of mineral transaction, which, in nature, is found upon a scale of grandeur and magnificence. They are nodules contained in the whinstone, porphyry, or basaltes of the Calton-hill, by Edinburgh; a body which is to be afterwards examined, when it will be found to have flowed, and to have been in fusion, by the operation of subterranean heat.

THIS evidence, though most conclusive with regard to the application of subterranean heat, as the means employed in bringing into fusion all the different substances with which strata may be found consolidated, is not directly a proof that strata had been consolidated by the fusion of their proper substance. It was necessary to see the general nature of the evidence, for the universal application of subterranean heat, in the fusion of every kind of mineral body. Now, that this has been done, we may give examples of strata consolidated without the introduction of foreign matter, merely by the softening or fusion of their own materials.

FOR this purpose, we may consider two different species of strata, such as are perfectly simple in their nature, of the most distinct substances, and whose origin is perfectly understood, consequently, whose subsequent changes may be reasoned upon with certainty and clearness. These are the siliceous and calcareous strata; and these are the two prevailing substances of the globe, all the rest being, in comparison of these, as nothing; for unless it be the bituminous or coal strata, there is hardly any other which does not necessarily contain more or less of one of other of these two substances. If, therefore, it can be shewn, that both of those two general strata have been consolidated by the simple fusion of their substance, no *desideratum* or doubt will remain, with regard to the nature of that operation which has been transacted at great depths of the earth, places to which all access is denied to mortal eyes.

WE are now to prove, *first*, That those strata have been consolidated by simple fusion; and, 2*dly*, That this operation is universal, in relation to the strata of the earth, as having produced all various degrees of solidity or hardness in these bodies.

I SHALL first remark, that a fortuitous collection of hard bodies, such as gravel and sand, can only touch in points and cannot, while in that hard state, be made to correspond so precisely to each other's shape as to consolidate the mass. But if these hard bodies should be softened in their substance, or brought into a certain degree of fusion, they might be adapted mutually to each other, and thus consolidate the open structure of the mass. Therefore, to prove the present point, we have but to exhibit specimens of siliceous and calcareous strata which have been evidently consolidated in this manner.

OF the first kind, great varieties occur in this country. It is, therefore, needless to describe these particularly. They are the consolidated strata of gravel and sand, often containing abundance of feld-spar, and thus graduating into granite; a body, in this respect, perfectly similar to the more regular strata which we now examine.

THE second kind, again, are not so common in this country, unless we consider the shells and coralline bodies in our limestones, as exhibiting the same example, which indeed they do. But I have a specimen of marble from Spain, which may be described, and which will afford the most satisfactory evidence of the fact in question.

THIS Spanish marble may be considered as a species of pudding-stone, being formed of calcareous gravel; a species of marble which, from Mr BOWLES'S Natural History, appears to be very common in Spain. The gravel of which this marble is composed, consists of fragments of other marbles of different kinds. Among these, are different species of *oolites* marble, some shell marbles, and some composed of a chalky substance, or of undistinguishable parts. But it appears, that all these different marbles had been consolidated or made hard, then broken into fragments, rolled and worn by attrition, and thus collected together, along with some sand or small siliceous bodies, into one mass. *Lastly*, This compound body is consolidated in such a manner as to give the most distinct evidence, that this had been executed by the operation of heat or simple fusion.

THE proof I have is this, That besides the general conformation of those hard bodies, so as to be perfectly adapted to each other's shape, there is, in some places, a mutual indentation of the different pieces of gravel into each other; an indentation which resembles perfectly that junction of the different bones of the *cranium*, called sutures, and which must have necessarily required a mixture of those bodies while in a soft or fluid state.

THIS appearance of indentation is, by no means, singular or limited to one particular specimen. I have several specimens of different marbles, in which fine examples of this species of mixture may be perceived. But in this particular case of the Spanish pudding-stone, where the mutual indentation is made between two pieces of hard stone, worn round by attrition, the softening or fusion of these two bodies is not simply rendered probable, but demonstrated.

HAVING thus proved, that those strata had been consolidated by simple fusion, as proposed, we now proceed to shew, that this mineral operation had been not only general, as being found in all regions of the globe, but universal, in consolidating our earth in all the various degrees, from loose and incoherent shells and sand, to the most solid bodies of the siliceous and calcareous substances.

TO exemplify this in the various collections and mixtures of sands, gravels, shells and corals, were endless and superfluous. I shall only take, for an example, one simple homogeneous body, in order to exhibit it in the various degrees of consolidation, from the state of simple incoherent earth to that of the most solid marble. It must be evident that this is chalk; naturally a soft calcareous earth, but which may be also found consolidated in every different degree.

THROUGH the middle of the isle of Wight, there runs a ridge of hills of indurated chalk. This ridge runs from the isle of Wight directly west into Dorsetshire, and goes by Corf-castle towards Dorchester, perhaps beyond that place. The sea has broke through this ridge at the west end of the isle of Wight, where columns of the indurated chalk remain, called the needles; the same appearance being found upon the opposite shore in Dorsetshire.

IN this field of chalk, we find every gradation of that soft earthy substance to the most consolidated body of this indurated ridge, which is not solid marble, but which has lost its chalky property, and has acquired a kind of stony hardness.

WE want only further to see this cretaceous substance in its most indurated and consolidated state; and this we have in the north of Ireland, not far from the Giants Causeway. I have examined cargoes of this limestone brought to the west of Scotland, and find the most perfect evidence of this body having been once a mass of chalk, which is now a solid marble.

THUS, if it is by means of fusion that the strata of the earth have been, in many places, consolidated, we must conclude, that all the degrees of consolidation, which are indefinite, have been brought about by the same means.

NOW, that all the strata of the mineral regions, which are those only now examined, have been consolidated in some degree, is a fact for which no proof can be offered here, but must be submitted to experience and enquiry; so far, however, as they shall be considered as consolidated in any degree, which they certainly are in general, we have investigated the means which had been employed in that mineral operation.

WE have now considered the concretions of particular bodies, and the general consolidation of strata; but it may be alleged, that there is a great part of the solid mass of this earth not properly comprehended among those bodies which have been thus proved to be consolidated by means of fusion. The body here alluded to is granite; a mass which is not generally stratified, and which, being a body perfectly solid, and forming some part in the structure of this earth, deserves to be considered.

THE nature of granite, as a part of the structure of the earth, is too intricate a subject to be here considered, where we only seek to prove the fusion of a substance from the evident marks which are to be observed in a body. We shall, therefore, only now consider one particular species of granite; and if this shall appear to have been in a fluid state of fusion, we may be allowed to extend this property to all the kind.

THE species now to be examined comes from the north country, about four or five miles west from Portsoy, on the road to Huntly. I have not been upon the spot, but am informed that this rock is immediately connected or continuous with the common granite of the country. This indeed appears in the specimens which I have got; for, in some of these, there is to be perceived a gradation from the regular to the irregular sort.

THIS rock may indeed by considered, in some respects, as a porphyry; for it has an evident ground, which is feld-spar, in its sparry state; and it is, in one view, distinctly maculated with quartz, which is transparent, but somewhat dark-coloured.[6]

CONSIDERED as a porphyry, this specimen is no less singular than as a granite. For, instead of a siliceous ground, maculated with the rhombic feld-spar, which is the common state of porphyry, the ground is uniformly crystallized, or a homogeneous regular fel-spar, maculated with the transparent siliceous substance. But as, besides the fel-spar and quartz, which are the constituent parts of the stone, there is also mica, in some places, it may, with propriety, be termed a granite.

THE singularity of this specimen consists, not in the nature or proportions of its constituent parts, but in the uniformity of the sparry ground, and the regular shape of the quartz mixture. This siliceous substance, viewed in one direction, or longitudinally, may be considered as columnar, prismatical, or continued in lines running nearly parallel. These columnar bodies of quartz are beautifully impressed with a figure on the sides, where they are in contact with the spar. This figure is that of furrows or channels, which are perfectly parallel, and run across the longitudinal direction of the quartz. This is represented in fig. 4. This striated figure is only seen when, by fracture, the quartz is separated from the contiguous spar.

BUT what I would here more particularly represent is, the transverse section of those longitudinal siliceous bodies. These are seen in fig. 1. 2. and 3. They have not only separately the forms of certain typographic characters, but collectively give the regular lineal appearance of types set in writing.

IT is evident from the inspection of this fossil, that the sparry and siliceous substances had been mixed together in a fluid state; and that the crystallization of the sparry substance, which is rhombic, had determined the regular structure of the quartz, at least in some directions.

THUS, the siliceous substance is to be considered as included in the spar, and as figured according to the laws of crystallization proper to the sparry ground; but the spar is also to be found included in the quartz. IT is not, indeed, always perfectly included or inclosed on all sides; but this is sometimes the case, or it appears so in the section. Fig. 5. 6. 7. 8. 9. and 10. are those cases magnified, and represent the different figured quartz inclosing the feld-spar. In one of them, the feld-spar, which is

---

[6] Plate II. fig.I. 2. 3.

contained within the quartz, contains also a small triangle of quartz, which it incloses. Now, it is not possible to conceive any other way in which those two substances, quartz and feld-spar, could be thus concreted, except by congelation from a fluid state, in which they have been mixed.

THERE is one thing more to be observed with regard to this curious species of granite. It is the different order or arrangement of the crystallization or internal structure of the feld-spar ground, in two contiguous parts of the same mass. This to be perceived in the polished surface of the stone, by means of the reflection of light.

THERE is a certain direction in which, viewing the stone, when the light falls with a proper obliquity, we see a luminous reflection from the internal parts of the stone. This arises from the reflecting surfaces of the sparry structure or minute cracks, all turned in one direction, consequently, giving that luminous appearance only in one point of view.

NOW, all the parts of the stone in which the figured quartz is directed in the same manner, or regularly placed in relation to each other, present that shining appearance to the eye at one time, or in the same point of direction. But there are parts of the mass, which, though immediately contiguous and properly continuous, have a different disposition of the figured quartz; and these two distinguished masses, in the same surface of the polished stone, give to the eye their shining appearance in very different directions. Fig. 3 shows two of those figured and shining masses, in the same plane or polished surface.

IT must be evident, that, as the crystallization of the sparry structure is the figuring cause of the quartz bodies, there must be observed a certain correspondency between those two things, the alinement (if I may be allowed the expression) of the quartz, and the shining of the sparry ground. It must also appear, that, at the time of congelation of the fluid spar, those two contiguous portions had been differently disposed in the crystallization of their substance. This is an observation which I have had frequent opportunities of making, with respect to masses of calcareous spar.

UPON the whole, therefore, whether we shall consider granite as a stratum or as an irregular mass, whether as a collection of several material, or as the separation of substances which had been mixed, there is sufficient evidence of this body having been consolidated by means of fusion, and in no other manner. WE are thus led to suppose, that the power of heat and operation of fusion must have been employed in consolidating strata of loose materials, which had been collected together and amassed at the bottom of the ocean. It will, therefore, be proper to consider, what are the appearances in consolidated strata that naturally should follow, on the one hand, from fluidity having been, in this manner, introduced by means of heat, and, on the other, from the interstices being filled by means of solution; that so we may compare appearances with the one and other of those two suppositions, in order to know that with which they may be only found consistent.

THE consolidation of strata with every different kind of substance was found to be inconsistent with the supposition, that aqueous solution had been the means employed for this purpose. This appearance, on the contrary, is perfectly consistent with the idea, that the fluidity of these bodies had been the effect of heat; for, whether we suppose the introduction of foreign matter into the porous mass of a stratum for its consolidation, or whether we shall suppose the materials of the mass acquiring a degree of softness, by means of which, together with an immense compression, the porous body might be rendered solid; the power of heat, as the cause of fluidity and vapour, is equally proper and perfectly competent. Here, therefore, appearances are as decidedly in favour of the last supposition, as they had been inconsistent with the first.

BUT if strata have been consolidated by means of aqueous solution, these masses should be found precisely in the same state as when they were originally deposited from the water. The perpendicular section of those masses might shew the compression of the bodies included in them, or of which they are composed; but the horizontal section could not contain any separation of the parts of the stratum from one another.

IF, again, strata have been consolidated by means of heat, acting in such a manner as to soften their substance, then, in cooling, they must have formed rents or separations of their substance, by the unequal degrees of contraction which the contiguous strata may have suffered. Here is a most decisive mark by which the present question must be determined.

THERE is not in nature any appearance more distinct than this of the perpendicular fissures and separations in strata. These are generally known to workmen by the terms of veins or backs and cutters; and there is no consolidated stratum that wants these appearances. Here is, therefore, a clear decision of the question, whether it has been by means of heat, or by means of aqueous solution, that collections of loose bodies at the bottom of the sea have been consolidated into the hardest rocks and most perfect marbles.

ERROR never can be consistent, nor can truth fail of having support from the accurate examination of every circumstance. It is not enough to have found appearances decisive of the question, with regard to the two suppositions which have been now considered. we may farther seek confirmation of that supposition which has been found alone consistent with appearances.

IF it be by means of heat and fusion that strata have been consolidated, then, in proportion to the degree of consolidation they have undergone from their original state, they should, *caeteris paribus*, abound more with separations in their mass. But this conclusion is found consistent with appearances. A stratum of porous sand-stone does not abound so much with veins and cutters as a similar stratum of marble, or even a similar stratum of stand-stone that is more consolidated. In proportion, therefore, as strata have been consolidated, they are in general intersected with veins and cutters; and in proportion as strata are deep in their perpendicular section, the veins are wide, and placed at greater distances. In like manner, when strata are thin, the veins are many, but proportionally narrow.

IT is thus, upon chemical principles, to be demonstrated, that all the solid strata of the globe have been condensed by means of heat, and hardened from a state of fusion. But this proposition is equally to be maintained from principles which are mechanical. The strata of the globe, besides being formed of earths, are composed of sand, of gravel, and fragments of hard bodies, all which may be considered as, in their nature, simple; but these strata are also found composed of bodies which are not simple, but are fragments of former strata, which had been consolidated, and afterwards were broken and worn by attrition, so as to be made gravel. Strata composed in this manner have been again consolidated; and now the question is, by what means?

IF strata composed of such various bodies had been consolidated, by any manner of concretion, from the fluidity of a dissolution, the hard and solid bodies must be found in their entire state, while the interstices between those constituent parts of the stratum are filled up. No partial fracture can be conceived as introduced into the middle of a solid mass of hard matter, without having been communicated from the surrounding parts. But such partial separations are found in the middle of those hard and solid masses; therefore, this compound body must have been consolidated by other means than that of concretion from a state of solution.

THE Spanish marble already described, as well as many consolidated strata of siliceous gravel, of which I have specimens, afford the clearest evidence of this fact. These hard bodies are perfectly united together, in forming the most solid mass; the contiguous parts of some of the rounded fragments are interlaced together, as has already been observed; and there are partial shrinkings of the mass forming veins, traversing several fragments, but perfectly filled with the sparry substance of the mass, and sometimes with parts of the stone distinctly floating in the transparent body of spar. Now, there is not, besides heat or fusion, any known power in nature by which these effects might be produced. But such effects are general to all consolidated masses, although not always so well illustrated in a cabinet specimen.

THUS we have discovered a truth that is confirmed by every appearance, so far as the nature of the subject now examined admits. We now return to the general operation, of forming continents of those materials which had been deposited at the bottom of the sea.

PART III.

Investigation of the Natural Operations employed in the Production of Land above the Surface of the Sea

WE seek to know that operation by means of which masses of loose materials, collected at the bottom of the sea, were raised above its surface, and transformed into solid land.

WE have found, that there is not in this globe (as a planet revolving in the solar system) any power or motion adapted to the purpose now in view; nor, were there such a power, could a mass of simply collected materials have continued any considerable time to resist the waves and currents natural to the sea, but must have been quickly carried away, and again deposited at the bottom of the ocean. But we have found, that there had been operations, natural to the bowels of this earth, by which those loose and unconnected materials have been cemented together, and consolidated into masses of great strength and hardness; those bodies are thus enabled to resist the force of waves and currents, and to preserve themselves, for a sufficient time, in their proper shape and place, as land above the general surface of the ocean. We now desire to know, how far those internal operations of the globe, by which solidity and stability are procured to the beds of loose materials, may have been also employed in raising up a continent of land, to remain above the surface of the sea.

THERE is nothing so proper for the erection of land above the level of the ocean, as an expansive power of sufficient force, applied directly under materials in the bottom of the sea, under a mass that is proper for the formation of land when thus erected. The question is not, how such a power may be procured; such a power has probably been employed. If, therefore, such a power should be consistent with that which we found had actually been employed in preparing the erected mass; or, if such a power is to be reasonable concluded as accompanying those operations which we have found natural to the globe, and situated in the very place where this expansive power appears to be required, we should thus be led to perceive, in the natural operations of the globe, a power as efficacious for the elevation of what had been at the bottom of the sea into the place of land, as it is perfect for the preparation of those materials to serve the purpose of their elevation.

IN opposition to this conclusion, it will not be allowed to allege, that we are ignorant how such a power might be exerted under the bottom of the ocean; for the present question is not, what had been the cause of heat, which has appeared to have been produced in that place; but, if this power of heat, which has certainly been exerted at the bottom of the ocean for consolidating strata, had been employed also for another purpose, that is, for raising those strata into the place of land.

WE may, perhaps, account for the elevation of land, by the same cause with that of the consolidation of strata, already investigated, without explaining the means employed by nature in procuring the power of heat, or shewing from what general source of action this particular power had been derived; but, by finding in subterranean heat a cause for any other change, besides the consolidation of porous or incoherent bodies, we shall generalize a fact, or extend our knowledge in the explanation of natural appearances.

THE power of heat for the expansion of bodies, is, so far as we know, unlimited; but by the expansion of bodies placed under the strata at the bottom of the sea, the elevation of those strata may be affected; and the question now to be resolved regards the actual exertion of this power of expansion, how far it is to be concluded as having been employed in the production of this earth above the level of the sea.

BEFORE attempting to resolve that question, it may be proper to observe, there has been exerted an extreme degree of heat below the strata formed at the bottom of the sea; and this is precisely the action of a power required for the elevation of those heated bodies into a higher place. Therefore, if there is no other way in which we may conceive this event to have been brought about, consistent with the present state of things, or what actually appears, we shall have a right to conclude, that such had been the order of procedure in natural things, and that the strata formed at the bottom of the sea had been elevated, as well as consolidated, by means of subterraneous heat.

THE consolidation of strata by means of fusion or the power of heat, has been concluded from the examination of nature, and from finding, that the present state of things is inconsistent with any other supposition. Now, again, we are considering the only power that may be conceived as capable of elevating strata from the bottom of the sea, and placing such a mass above the surface of the water. It is a truth unquestionable, that what had been originally at the bottom of the sea, is at present the highest of our land. In explaining this appearance, therefore, no other alternative is left, but either to suppose strata elevated by the power of heat above the level of the present sea, or the surface of the ocean reduced many miles below the height at which it had subsisted during the collection and induration of the land which we inhabit.

NOW, if, on the one hand, we are to suppose no general power of subterraneous fire or heat, we leave to our theory no means for the retreat of the sea, or the lowering of its surface; if, on the other hand, we are to allow the general power of subterraneous heat, we cannot have much difficulty in supposing, either the surface of the sea to have subsided, or the bottom of the ocean, in certain parts, to have been raised by a subterranean power above the level of the surface, according as appearances shall be found to require the one or other of those conclusions. Here, therefore, we are again remitted to the history of nature, in order to find matter of fact by which this question may be properly decided.

IF the present land had been discovered by the subsiding of the waters, there has not been a former land, from whence materials had been procured for the construction of the present, when at the bottom of the sea; for there is no vestige remaining of that land, the whole land of the present earth having been formed evidently at the bottom of the sea. Neither could the natural productions of the sea have been accumulated, in the shape in which we now find them, on the surface of this earth; for how should the Alps and Andes have been formed within the sea from the natural productions of the water? Consequently, this is a supposition inconsistent with every natural appearance.

THE supposition, therefore, of the subsidence of the former ocean, for the purpose of discovering the present land, is beset with more difficulty than the simple erection of the bottom of the former ocean; for, *first*, There is a place to provide for the retirement of the waters of the ocean; and, 2*dly*, There is required a work of equal magnitude; that is, the swallowing up of that former continent, which had procured the materials of the present land. ON the one hand, the subsiding of the surface of the ocean would but make the former land appear the higher; and, on the other, the sinking the body of the former land into the solid globe, so as swallow up the greater art of the ocean after it, if not a natural impossibility, would be a least a superfluous exertion of the powers of nature. Such an operation as this would discover as little wisdom in the end elected, as in the means appropriated to that end; for , if the land be not wafted and worn away in the natural operations of the globe, why make such a convulsion in the world in order to renew the land? If, again, the land naturally decays, why employ so extraordinary a power, in order to hide a former continent of land, and puzzle man?

LET us now consider how far the other proposition, of strata being elevated by the power of heat above the level of the sea, may be confirmed from the examination of natural appearances.

THE strata formed at the bottom of the ocean are necessarily horizontal in their position, or nearly so, and continuous in their horizontal direction or extent. They may change, and gradually assume the nature of each other, so far as concerns the materials of which they are formed; but there cannot be any sudden change, fracture or displacement naturally in the body of a stratum. But, if these strata are cemented by the heat of fusion, and erected with an expansive power acting below, we may expect to find every species of fracture, dislocation and contortion, in those bodies, and every degree of departure from a horizontal towards a vertical position.

THE strata of the globe are actually found in every possible position: For from horizontal, they are frequently found vertical; from continuous, they are broken and separated in every possible direction; and, from a plane, they are bent and doubled. It is impossible that they could have originally been formed, by the known laws of nature, in their present state and position; and the power that has been necessarily required for their change, has not been inferior to that which might have been required for their elevation from the place in which they had been formed.

IN this case, natural appearances are not anomalous. They are, indeed, infinitely various, as they ought to be, according to the rule; but all those varieties in appearances conspire to prove one general truth, *viz*. That all which we see had been originally composed according to certain principles, established in the constitution of the terraqueous globe; and that those regular compositions had been afterwards greatly changed by the operations of another power, which had introduced apparent confusion among things first formed in order and by rule.

IT is concerning the operation of this second power that we are now enquiring; and here the apparent irregularity and disorder of the mineral regions are as instructive, with regard to what had been transacted in a former period of time, as the order and regularity of those same regions are conclusive, in relation to the place in which a former state of things had produced that which, in its changed state, we now perceive.

WE are now to conclude, that the land on which we dwell had been elevated from a lower situation by the same agent which had been employed in consolidating the strata, in giving them stability, and preparing them for the purpose of the living world. This agent is matter actuated by extreme heat, and expanded with amazing force.

IF this has been the case, it will reasonable to expect, that some of the expanded matter might be found condensed in the bodies which have been heated by that igneous vapour; and that matter, foreign to the strata, may have been thus introduced into the fractures and separations of those indurated masses.

WE have but to open our eyes to be convinced of this truth. Look into the sources of our mineral treasures; ask the miner from whence has come the metal into his veins? Not from the earth or air above, not from the strata which the vein traverses; these do not contain one atom of the minerals now considered: There is but one place from whence these minerals may have come; this is, the bowels of the earth, the place of power and expansion, the place from whence must have proceeded that intense heat by which loose materials have been consolidated into rocks, as well as that enormous force by which the regular strata have been broken and displaced.

OUR attention is here peculiarly called upon, where we have the opportunity of examining those mineral bodies, which have immediately proceeded from the unknown region, that place of power and energy which we want to explore; for, if such is the system of the earth, that materials are first deposited at the bottom of the ocean, there to be prepared in a certain manner, in order to acquire solidity, and then to be elevated into the proper place of land, these mineral veins, which contain matter absolutely foreign to the surface of the earth, afford the most authentic information with regard to the operations which we want to understand. It is these veins which we are to consider as, in some measure, the continuation of that mineral region, which lies necessarily out of all possible reach of our examination. It is, therefore, peculiarly interesting to know the state in which things are to be found in this place, which may be considered as intermediate between the solid land, upon the one hand, and the unknown regions of the earth, upon the other.

WE are now to examine those mineral veins; and these may be considered, first, in relation to their form, independent of their substance or particular contents; and, secondly, in relation to the contained bodies, independent of their form.

IN examining consolidated strata, we remarked veins and cutters as a proof of the means by which those bodies had been consolidated. In that case, the formation of these veins is a regulated process, determined by the degree of fusion, and the circumstances of condensation or refrigeration. In respect of these, the mineral veins now to be examined are anomalous. They are; but we know not why or how. We see the effect; but, in that effect, we do not see the cause. We can say, negatively, that the cause of mineral veins is not that by which the veins and fissures of consolidated strata have been formed; consequently, that is not the measured contraction and regulated condensation of the consolidated land which has formed those general mineral veins; however, veins, similar in many respects, have been formed by the co-operation of this cause.

HAVING thus taken a view of the evident distinction between the veins or contractions that are particular to the consolidated body in which they are found, and those more general veins which are not limited to that cause, we may now consider what is general in the subject, or what is universal in these effects of which we wish to investigate the cause.

THE event of highest generalization or universality, in the form of those mineral veins, is fracture and dislocation. It is not, like that of the veins of strata, simple separation and measured contraction; it is violent fracture and unlimited dislocation. In the one case, the forming cause is in the body which is separated; for, after the body had been actuated by heat, it is by the reaction of the proper matter of the body, that the chasm which constitutes the vein is formed. In the other case, again, the cause is extrinsic in relation to the body in which the chasm is formed. There has been the most violent fracture and divulsion; but the cause is still to seek; and it appears not in the vein; for it is not every fracture and dislocation of the solid body of our earth, in which minerals, or the proper substances of mineral veins, are found.

WE are now examining matter of fact, real effects, from whence we would investigate the nature of certain events which do not now appear. Of these, two kinds occur; one which has acted in relation to the hardness and solidity, or the natural constitution of the body; the other, to its shape or local situation. The first has been already considered; the last is now the subject of enquiry.

BUT, in examining those natural appearances, we find two different kinds of veins; the one necessarily connected with the consolidating causes; the other with that cause of which we now particularly enquire. For, in those great mineral veins, violent fracture and dislocation is the principle; but there is no other principle upon which strata, or masses formed at the bottom of the sea, can be placed at a height above its surface. Hence, in those two different operations, for forming mineral veins, and erecting strata form a lower to a higher place, the principle is the same; for neither can be done without violent fracture and dislocation.

WE now only want to know, how far it is by the same power, as well as upon the same principle, that those two operations have been made. An expansive force, acting from below, is the power most proper for erecting masses; but whether it is a power of the same nature with that which has been employed in forming mineral veins, will best appear in knowing the nature of their contents. These, therefore, may be now considered.

EVERY species of fracture, and every degree of dislocation and contortion, may be received in the form of mineral veins; and there is no other general principle to be observed in examining their form. But, in examining their contents, some other principle may appear, so far as, to the dislocating power or force, there may be superadded matter, by which something in relation to the nature of the power may be known. If, for example, a tree or a rock shall be found simply split asunder, although there be no doubt with regard to some power having been applied in order to produce the effect, yet we are left merely to conjecture at the power. But when wedges of wood or iron, or frozen water, should be found lodged in the cleft, we might be enabled, from this appearance, to form a certain judgment with regard to the nature of the power which had been applied. This is the case with mineral veins. We find them containing matter, which indicates a cause; and every information in this case is interesting to the theory.

THE substances contained in mineral veins are precisely the same with those which, in the former part of this paper, we have considered as being made instrumental in the consolidation of strata; and they are found in every species of mixture and concretion.

BUT, besides this evidence for the exertion of extreme heat, in that process by which those veins were filled, there is another important observation to be gathered from the inspection of this subject. There appears to have been a great mechanical power employed in the filling of these veins, as well as that necessarily required in making the first fracture and divulsion.

THIS appears from the order of the contents, or filling of these veins, which is a thing often observed to be various and successive. But what it is chiefly now in view to illustrate, is that immense force which is manifested in the fracture and dispersion of the solid contents which had formerly filled those veins. Here we find fragments of rock and spar floating in the body of a vein filled with metallic substances; there, again, we see the various fragments of metallic masses floating in the sparry and siliceous contents.

ONE thing is demonstrable from the inspection of the veins and their contents; this is, the successive irruptions of those fluid substances breaking the solid bodies which they meet, and floating those fragments of the broken bodies in the vein. It is very common to see three successive series of those operations; and all this may be perceived in a small fragment of stone, which a man of science may examine in his closet, often better than descending to the mine, where all the examples are found on an enlarged scale.

LET us now consider what power would be required to force up, from the most unfathomable depth of the ocean, to the Andes or the Alps, a column for fluid metal and of stone. This power cannot be much less than that required to elevate the highest land upon the globe. Whether, therefore, we shall consider the general veins as having been filled by mineral steams, or by fluid minerals, an elevating power of immense force is still required, in order to form as well as fill those veins. But such a power acting under the consolidated masses at the bottom of the sea, is the only natural means for making those masses land.

IF such have been the operations that are necessary for the production of this land; and if these operations are natural to the globe of this earth, as being the effect of wisdom in its contrivance, we shall have reason to look for the actual manifestation of this truth in the phaenomena of nature, or those appearances which more immediately discover the actual cause in the perceived effect.

TO see the evidence of marble, a body that is solid, having been formed of loose materials collected at the bottom of the sea, is not always easy, although it may be made abundantly plain; and to be convinced that this calcareous stone, which calcines so easily in our fires, should have been brought into fusion by subterraneous heat, without suffering calcination, must require a chain of reasoning which ever one is not able to attain. But when fire bursts forth from the bottom of the sea, and when the land is heaved up and down, so as to demolish cities in an instant, and split asunder rocks and solid mountains, there is nobody but must see in this a power, which may be sufficient to accomplish every view of nature in erecting land, as it is situated in the place most advantageous for that purpose. THE only question, therefore, which it concerns us to decide at present, is, whether those operations of extreme heat, and violent mechanic force, be only in the system as a matter of accident; or if, on the contrary, they are operations natural to the globe, and necessary in the production of such land as this which we inhabit. The answer to this is plain: these operations of the globe, remain at present with undiminished activity, or in the fulness of their power.

A stream of melted lava flows from the sides of Mount Ætna. Here a column of weighty matter raised an immense height above the level of the sea, and rocks of an enormous size are projected from its orifice some miles into the air. Every one acknowledges that here is the liquefying power and expansive force of subterranean fire, or violent heat. But that Sicily itself had been raised from the bottom of the ocean, and that the marble called Sicilian Jasper, had its solidity upon the same principle with the lava, would stumble many a naturalist to acknowledge. Nevertheless, I have in my possession a table of this marble, from which it is demonstrable, that this calcareous stone had flowed, and been in such a state of fusion and fluidity as lava.

HERE is a comparison formed of two mineral substances, to which it is of the highest importance to attend. The solidity and present state of the one of these is commonly thought to be that of water. This, however, is not the case. The immediate state and condition of both these bodies is now to be considered as equally the effect of fire or heat. The reason of our forming such a different judgment with regard to these two subjects is this; we see, in the one case, the more immediate connection of the cause and the effect, while, in the other, we have only the effects from whence we are in science to investigate the cause. BUT, if it were necessary always to see this immediate connection, in order to acknowledge the operation of a power which, at present, is extinguished in the effect, we should lose the benefit of science, or general principles, from whence particulars may be deduced, and we should be able to reason no better that the brute. Man is made for science; he reasons from effects to causes, and from causes to effects; but he does not always reason without error. In reasoning, therefore, from appearances which are particular, care must be taken how we generalize; we should be cautious not to attribute to nature, laws which may perhaps be only of our own invention.

THE immediate question now before us is not, if the subterraneous fire, or elevating power, which we perceive sometimes as operating with such energy, be the consolidating cause of strata formed at the bottom of the sea; nor, if that power be the means of making land appear above the general surface of the water; for, though this be the end we want to arrive at ultimately, the question at present in agitation respects the laws of nature, or the generality of particular appearances.

HAS the globe within it such an active power as fits it for the renovation of that part of its constitution which may be subject to decay? Are those powerful operations of fire, or subterraneous heat, which so often have filled us with terror and astonishment, to be considered as having always been? Are they to be concluded as proper to every part upon the globe, and as continual in the system of this earth? If these points in question shall be decided in the affirmative, we can be at no loss in ascertaining the power which has consolidated strata, nor in explaining the present situation of those bodies, which had their origin at the bottom of the sea. This, therefore, should be the object of our pursuit; and, in order to have demonstration in a case of physical enquiry, we must again have recourse to the book of nature. THE general tendency of heat is to produce fluidity and softness; as that of cold is, on the contrary, to harden soft and fluid bodies. But this softening power of heat is not uniform in its nature; it is made to act with very different effect, according to the nature of the substance to which it is applied. We are but limited in the art of increasing the heat or the cold of bodies; we find, however, extreme difference in their substances with respect to fusibility.

A FUSIBLE substance, or mineral composition in a fluid state, is emitted from those places of the earth at which subterraneous fire and expansive force are manifested in those eruptive operations. In examining these emitted bodies, men of science find a character for such productions, in generalizing the substance, and understanding the natural constitution of those bodies. It is in this manner, that such a person, finding a piece of lava in any place of the earth, says with certainty, Here is a stone which had congealed from a melted state.

HAVING thus found a distinguishing character for those fused substances called, in general, lavas, and having the most visible marks for that which had been actually a volcano, naturalists, in examining different countries, have discovered the most undoubted proofs of many ancient volcanos, which had not been before suspected. Thus, volcanos will appear to be not a matter of accident, or as only happening in an particular place, they are general to the globe, so far as there is no place upon the earth that may not have an eruption of this kind; although it is by no means necessary for every place to have had those eruptions.

VOLCANOS are natural to the globe, as general operations; but we are not to consider nature as having a burning mountain for an end in her intention, or as a principal purpose in the general system of this world. The end of nature in placing an internal fire or power of heat, and a force of irresistible expansion, in the body of this earth, is to consolidate the sediment collected at the bottom of the sea, and to form thereof a mass of permanent land above the level of the ocean, for the purpose of maintaining plants and animals. The power appointed for this purpose is, as on all other occasions, where the operation is important, and where there is any danger of a shortcoming, wisely provided in abundance; and there are contrived means for disposing of the redundancy. These, in the present case, are our volcanos.

A VOLCANO is not made on purpose to frighten superstitious people into fits of piety and devotion, nor to overwhelm devoted cities with destruction; a volcano should be considered as a spiracle to the subterranean furnace, in order to prevent the unnecessary elevation of land, and fatal effects of earthquakes; and we may rest assured, that they, in general, wisely answer the end of their intention, without being in themselves an end, for which nature had exerted such amazing power and excellent contrivance.

LET us take a view of the most elevated places of the earth; if the present theory is just, it is there that we should find volcanos. But is not this the case? There are volcanos in the Andes; and round the Alps we find many volcanos, which are in France upon the one side, and in Germany upon the other, as well as upon the Italian side, where Vesuvius still continues to exhibit violent eruptions.

IT is not meant to allege, that it is only upon the summit of a continent volcanos should appear. Subterraneous fire has sometimes made its appearance in bursting from the bottom of the sea. But, even in this last case, land was raised from the bottom of the sea, before the eruption made its exit into the atmosphere. It must also be evident, that, in this case of the new island near Santorini, had the expansive power been retained, instead of being discharged. much more land might have been raised above the level of the ocean. NOW, the eruption of that elastic force through the bottom of the sea, may be considered as a waste of power in the operations of the globe, where the elevation of the indurated strata is an object in the exertion of that power; whereas, in the centre of a continent sufficiently elevated above the level of the sea, the eruption of that fiery vapour calculated to elevate the land, while it may occasionally destroy the habitations of a few, provides for the security and quiet possession of the many.

IN order to see the wisdom of this contrivance, let us consider the two extreme places at which this eruption of ignited matter may be performed. These are, on the one hand, within a continent of land, and, on the other, at the bottom of the ocean. In the one case, the free eruption of the expanding power should be permitted; because the purpose for which it had been calculated to exist, has been accomplished. In the other, again, the free eruption of that powerful matter should be repressed; because there is reserved for that power much of another operation in that place. But, according to the wise constitution of things, this must necessarily happen. The eruption of the fiery vapour from volcanos on the continent or land, is interrupted only occasionally, by the melted bodies flowing in the subterraneous chimney; whereas, at the bottom of the ocean, the contact of the water necessarily tends to close the orifice, by accumulating condensed matter upon the weakest place.

IF this be a just theory of the natural operations of the globe, we shall have reason to expect, that great quantities of this melted matter or fusible substance may be found in form of lava, among the strata of the earth, where there are no visible marks of any volcano, or burning mountain, having existed. Here, therefore, is an important point to be determined; for, if it shall appear, that much of this melted matter, analogous to lave, has been forced to flow among the strata which had been formed at the bottom of the sea, and now are found forming dry land above its surface, it will be allowed, that we have discovered the secret operations of nature concocting future land, as well as those by which the present habitable earth had been produced from the bottom of the abyss. Here, therefore, we shall at present rest the argument, with endeavouring to shew that such is actually the case.

IT appears from CRONSTEDT's Mineralogy, that the rockstone, called trap by the Swedes, the amygdaloides and the schwarts-stein of the Germans, are the same with the whinstone of this country. This is also confirmed by specimens from Sweden, sent me by my friend Dr GAHN. Whatever, therefore, shall be ascertained with regard to our whinstone, may be so far generalised or extended to the countries of Norway, Sweden, and Germany.

THE whinstone of Scotland is also the same with the toadstone of Derbyshire, which is of the amygdaloides species; it is also the same with the ragstone of the south of Staffordshire, which is a simple whinstone, or perfect trap. England, therefore, must be included in this great space of land, the mineral operations of which we explore; and also Ireland, of which the Giants Causeway, and many others, are sufficient proof.

IN the south of Scotland, there is a ridge of hills, which extends from the west side of the island in Galloway to the east side in Berwickshire, composed of granite, of schistus, and of siliceous strata. The Grampians on the north, again, form another range of mountains of the same kind; and between these two great fields of broken, tumbled and distorted strata, there lies a field of lesser hardness and consolidation, in general; but a filed in which there is a great manifestation of subterraneous fire, and of exerted force.

THE strata in this space consist, in general, of sandstone, coal, limestone or marble, ironstone, and marl or argillaceous strata, with strata of analogous bodies, and the various compositions of these. But what is to the present purpose is this, that, through all this space, there are interspersed immense quantities of whinstone; a body which is to be distinguished as very different from lava; and now the disposition of this whinstone is to be considered.

SOMETIMES it is found in an irregular mass or mountain, as Mr CRONSTEDT has properly observed; but he has also said, that this is not the case in general. His words are: "It is oftener found in form of veins in mountains or another kind, running commonly in a serpentine manner, contrary or across to the direction of the rock itself."

THE origin of this form, in which the trap or whinstone appears, is most evident to inspection, when we consider that this solid body has been in a fluid state, and introduced, in that state, among strata which preserved their proper form. The strata appear to have been broken, and the two correspondent parts of those strata are separated to admit the flowing mass of whinstone.

A FINE example of this kind may be seen upon the south side of the Earn, on the road to Crief. It is twenty-four yards wide, stands perpendicular, and appears many feet above the surface of the ground. It runs from that eastward, and would seem to be the same with that which crosses the river Tay, in forming Campsy-lin above Stanley, as a lesser one of the same kind does below it. If have seen it at Lednoc upon the Ammon, where it forms a cascade in that river, about five or six miles west of Campsy-lin. It appears to run from the Tay east through Strathmore, so that it may be considered as having been traced for twenty or thirty miles, and westwards to Drummond castle, perhaps much farther.

TWO small veins of the same kind, only two or three feet wide, may be seen in the bed of the Water of Leith, traversing the horizontal strata, the one is above St BERNARD's well, the other immediately below it. But, more particularly, in the shire of Ayr, to the north of Irvine, there are to be seen upon the coast, between that and Scarmorly, in the space of about twenty miles, more than twenty or thirty such dykes (as they are called) of whinstone. Some of them are of a great thickness; and, in some places, there is perceived a short one, running at right angles, and communicating with the other two that run parallel.

THERE is in this country, and in Derbyshire,[7] another regular appearance of this stone, which CRONSTEDT has not mentioned. In this case, the strata are not broken in order to have the whinstone introduced, they are separated, and the whinstone is interjected in form of strata, having various degrees of regularity, and being of different thickness. On the south side of Edinburgh, I have seen, in little more than the space of a mile from east to west, nine or ten masses of whinstone interjected among the strata. These masses of whinstone are from three or four to an hundred feet thick, running parallel in planes inclined to the horizon, and forming with it an angle of about twenty or thirty degrees, as may be seen at all times in the hill of Salisbury Craggs.

HAVING thus described these masses, which have flowed by means of heat among the strata of the globe, strata which had been formed by subsidence at the bottom of the sea, it will now be proper to examine the difference that subsists between these subterraneous lavas, as they may be termed, and the analogous bodies, which are proper lavas, in having issued out of a volcano.[8] THERE can be no doubt that these two

---

[7] See Mr WHITEHURST'S Theory of the Earth.

[8] The Chevalier de Dolomieu, in his accurate examination of Ætna and the Lipari islands, has very well observed the distinction of these two different species of lavas; but without seeming to know the principle upon which this essential difference depends. No bias of system, therefore, can here be supposed as perverting the Chevalier's view, in taking those observations; and these are interesting to the present theory, as corresponding perfectly with the facts from whence it has been formed. It will be proper to give the account of these in his own words. La zeolite est tres-commune dans certains laves de l'Ethna; il feroit peut être possible d'y en recontrer des morceaux aussi gros que ceux que fournit l'isle de Ferroé. Quoique cette substance semble ici appartenir aux laves, je ne dirai cependant point que

different species of bodies have had the same origin, and that they are composed of the same materials nearly; but from the different circumstances of their production, there is formed a character to these bodies, by which they may be perfectly distinguished. The difference of those circumstances, consists in this; the one has been emitted to the atmosphere in its fluid state, the other only came to be exposed to the light in a long course of time, after it had congealed under the compression of an immense load of earth, and after certain operations, proper to the mineral regions, had been exercised upon the indurated mass. This is the cause of the difference between those erupted lavas, and our whinstone, toadstone, and the Swedish trap, which may be termed subterraneous lava. The visible effects of those different operations may now be mentioned.

IN the erupted lavas, those substances which are subject to calcine and vitrify in our fires, suffer similar changes, when delivered from a compression which has rendered them fixed, though in an extremely heated state. Thus, a lava in which there is much calcareous spar, when it comes to be exposed to the atmosphere, or delivered from the compressing force of its confinement, effervesces by the explosion of its fixed air; the calcareous earth, at the same time, vitrifies with the other substances: hence such violent ebullition in volcanos, and hence the emission of so much pumice-stone and ashes, which are of the same nature.

IN the body of our whinstone, on the contrary, there is now mark of calcination or vitrification. We frequently find in it much calcareous spar, or the *terra calcarea aerata*, which had been in a melted state by heat, and had been crystallized by congelation into a sparry form. Such is the *lapis amygdaloides*, and many of our whinstone rocks, which contain pebbles crystallized and variously figured, both calcareous, siliceous, and of a mixture in which both these substances form distinct parts. The specimens of this kind, which I have from the whinstone or porphyry rock of the Calton-hill, exhibit every species of mineral operation, in

---

toutes les zeolites soient volcaniques, ou unies à des matieres volcaniques; celles que l'on trouve en Allemagne sont, dit-on, dans des circonstances differentes; mais je doit annoncer que je n'ai trouvé cette substance en Sicile, que dans les seules laves qui evidemment ont coulé dans la mer, et qui ont été recouvertes par ses eaux. La zeolite des laves n'est point une dejection volcanique, ni une production du feu, ni même un matiere que les laves aient enveloppée lorsequ'elles etoient fluides; elle est le resultat d'une operation et d'une combinaison posterieure, auxquelles les eaux de la mer ont concouru. Les laves qui n'ont pas été submergées, n'en contiennent jamais. J'ai trouvé ces observation si constantes, que partout où je rencontrois de la zeolite, j'étois sûr de trouver d'autres preuves de submersion, et partout où je voyois des laves recouvertes des dépôts de l'eau, j'étois sûr de trouver de la zeolite, et un de ces faits m'a toujours indiqué l'autre. Je me suis servi avec succes de cette observation pour diriger mes recherches, et pour connoitre l'antiquité des laves. *Mineralogie de Volcans, par M. Faujas de Saint-Fond.* Here would appear to be the distinction of subterraneous lava, in which zeolite and calcareous spar may be found, and that which has flowed from a volcano, in which neither of these are ever observed.

forming jasper, figured agate, and marble; and they demonstrate, that this had been performed by heat or fusion.

I DO not mean to say, that this demonstration is direct; it is conditional, and proceeds upon the supposition, that the basaltic or porphyry rock, in which those specimens are found, is a body which had been in a melted state. Now, this is a supposition for which I have abundance of evidence, were it required; but naturalists are now sufficiently disposed to admit that proposition; they even draw conclusions from this fact, which, I think, they are not sufficiently warranted in doing; that is, from this appearance, they infer the former existence of volcanos in those places. For my part, though I have made those most strict examination, I never saw and vestige of such an event. That there are, in other countries, evident marks of volcanos which have been long extinguished, is unquestionably true; but naturalists, imagining that there are no other marks of subterraneous fire and fusion, except in the production of a lava, attribute to a volcano, as a cause, these effects, which only indicate the exertion of that power which might have been the cause of a volcano.

IF the theory now given be just, a rock of marble is no less a mark of subterraneous fire and fusion, than that of the basaltes; and the flowing of basaltic streams among strata broken and displaced, affords the most satisfactory evidence of those operations by which the body of our land had been elevated above the surface of the sea; but it gives no proof that the eruptive force of mineral vapours had been discharged in a burning mountain. Now, this discharge is essential in the proper idea of a volcano.

BESIDES this internal mark of an unerupted lava in the substance of the stone or body of the flowing mass, there are others which belong to it in common with all other mineral strata, consolidated by subterraneous fire, and changed from the place of their original formation; this is, the being broken and dislocated, and having veins of foreign matter formed in their separations and contractions.

IF these are mineral operations, proper to the lower regions of the earth, and exerted upon bodies under immense compression, such things will be sometimes found in the unerupted lavas, as well as in the contiguous bodies with which they are associated. If, on the contrary, these are operations proper to the surface of the earth, where the dissolving power of water and air take place, and where certain stalactical and ferruginous concretions are produced by these means; then, in erupted lavas, we should find mineral concretions, which concretions should be denied to bodies which had been consolidated at the bottom of the sea; that is to say, where, without the operation of subterraneous fire, no changes of that kind could have taken place, as has already been observed. But in the unerupted species of lava, that is to say, in our whinstone, every species of mineral appearance is occasionally to be found. Let those who have the opportunity to examine, say, what are to be found in proper lavas, that is, those of the erupted kind. Sir WILLIAM HAMILTON informed me, when I shewed him those mineral veins and spars in our whinstone, that he had never observed the like in lavas.

WE have now formed some conclusions with regard to the nature and production of those parts of the land of this globe which we have had the means of examining perfectly; but form the accounts of travellers, and from the specimens which are brought to us from distant parts, we have reason to believe, that all the rest of the earth is of the same nature with that which has been now considered. The great masses of the earth are the same every where; and all the different species of earths, of rocks or stone, which have as yet appeared, are to be found in the little space of this our island.

IT is true, that there are peculiar productions in the mineral kingdom which are rare, as being found only in few places; but these things are merely accidental in relation to the land, for they belong in property to those parts of the mineral region which we never see. Such are, the diamond of the east, the platina of the west, and the tin of Cornwall, Germany, and Sumatra. Gold and silver, though found in many countries, do not appear to be immediately necessary in the production of a habitable country. Iron, again, is universal in the operations of the globe, and is found often in that profusion which equals its utility. Between these two extremes, we find all other minerals, that is to say, here and there in moderate quantity, and apparently in some proportion to their use. But all these substances are to be considered as the vapours of the mineral regions, condensed occasionally in the crevices of the land; and it is only the rocks and strata (in which those mineral veins are found) that are now examined with regard to their original composition at the bottom of the sea, as well as to that operation by which those bodies had been indurated in their substance, and elevated from the place in which they had been formed.

THUS, we have sufficient reason to believe, that, in knowing the construction of the land in Europe, we know the constitution of the land in every part of the globe. Therefore, we may proceed to form general conclusions, from the knowledge of the mineral region, thus acquired in studying those parts which are seen.

HAVING thus found, *first*, That the consolidated and indurated masses of our strata had suffered the effects of violent heat and fusion; 2*dly*, That those strata, which had been formed in a regular manner at the bottom of the sea, have been violently bended, broken and removed from their original place and situation; and, *lastly*, Having now found the most indubitable proof, that the melting, breaking, and removing power of subterraneous fire, has been actually exerted upon this land which we examine, we cannot hesitate in ascribing these operations as a cause to those effects which are exposed to our view. Now, these may be considered as consisting the solid state and present situation of those stratified bodies, originally formed by subsidence in the ocean;

appearances which cannot, in reason, be ascribed to any other cause, and which, upon this principle, are perfectly explained.

IT is not meant to specify every particular in the means employed by nature for the elevation of our land. It is sufficient to have shewn, that there is, in nature, means employed for the consolidation of strata, formed originally of loose and incoherent materials; and that those same means have also been employed in changing the place and situation of those strata. But how describe an operation which man cannot have any opportunity of perceiving? Or how imagine that, for which, perhaps, there are not proper data to be found? We only know, that the land is raised by a power which has for principle subterraneous heat; but how that land is preserved in its elevated station, is a subject in which we have not even the means to form conjecture; at least, we ought to be cautious how we indulge conjecture in a subject where no means occur for trying that which is but supposition.

WE now proceed, from the facts which have been properly established, to reason with regard to the duration of this globe, or the general view of its operations, as a living world, maintaining plants and animals.

PART IV.

System of Decay and Renovation observed in the Earth

PHILOSOPHERS observing an apparent disorder and confusion in the solid parts of this globe, have been led to conclude, that there formerly existed a more regular and uniform state, in the constitution of this earth; that there had happened some destructive change; and that the original structure of the earth had been broken and disturbed by some violent operation, whether natural, or from a supernatural cause. Now, all these appearances, from which conclusions of this kind have been formed, find the most perfect explanation in the theory which we have been endeavouring to establish; for they are the facts from whence we have reasoned, in discovering the nature and constitution of this earth: therefore, there is no occasion for having recourse to any unnatural supposition of evil, to any destructive accident in nature, or to the agency of any preternatural cause, in explaining that which actually exists. IT is necessary for a living or inhabited world, that this should consist of land and water. It is also necessary, that the land should be solid and stable, resisting, with great power, the violent efforts of the ocean; and, at the same time, that this solid land should be resolved by the influence of the sun and atmosphere, so as to decay, and thus become a soil for vegetation. But these general intentions are perfectly fulfilled in the constitution of our earth, which has been now investigated. This great body being formed of different mixed masses, having various degrees of hardness and solubility, proper soil for plants is supplied from the gradual resolution of the solid parts; fertility in those soils arises from the mixture of different elementary substances; and stability is procured to that vegetable world, by the induration of certain bodies, those rocks and stones, which protect the softer masses of clay and soil.

IN this manner, also, will easily be explained those natural appearances which diversify the surface of the earth for the use of plants and animals, and those objects which beautify the face of nature for the contemplation of mankind. Such are, the distinctions of mountains and valleys, of lakes and river, of dry barren desarts and rich watered plains, of rocks which stand apparently unimpaired by the lapse of time, and sands with fluctuate with winds and tides. All these are the effects of steady causes; each of these has its proper purpose in the system of the earth; and in that system is contained another, which is that of living growing bodies, and of animated beings.

BUT, besides this, man, the intellectual being, has, in this subject of the mineral kingdom, the means of gratifying the desire of knowledge, a faculty by which he is distinguished from the animal, and by which he improves his mind in knowing causes. Man is not satisfied, like the brute, in seeing things which are; he seeks to know how things have been, and what they are to be. It is with pleasure that he observes order and regularity in the works of nature, instead of being disgusted with disorder and confusion; and he is made happy from the appearance of wisdom and benevolence in the design, instead of being left to suspect in the Author of nature, any of that imperfection which he finds in himself.

LET us now take a view of that system of mineral oecomony, in which may be perceived every mark of order and design, of provident wisdom and benevolence.

WE have been endeavouring to prove, that all the continents and island of this globe had been raised above the surface of the ocean; we have also aimed at pointing out the cause of this translation of matter, as well as of the general solidity of that which is raised to our view; but however this theory shall be received, no person of observation can entertain a doubt, that all, or almost all we see of this earth, had been originally formed at the bottom of the sea. We have now another object in our view; this is to investigate the operations of the globe, at the time that the foundation of this land was laying in the waters of the ocean, and to trace the existence and the nature of things, before the present land appeared above the surface of the waters. We should thus acquire some knowledge of the system according to which this world is ruled, both in its preservation and production; and we might be thus enabled to judge, how far the mineral system of the world shall appear to be contrived with all the wisdom, which is so manifest in what are termed the animal and vegetable kingdoms.

IT must not be imagined that this undertaking is a thing unreasonable in its nature; or that it is a work necessarily beset with any unsurmountable difficulty; for, however imperfectly we may fulfil this end proposed, yet, so far as it is to natural causes that are to be ascribed the operations of former time, and so far as, from the present state of things, or knowledge of natural history, we have it in our power to reason from effect to cause, there are, in the constitution of the world, which we now examine, certain means to read the annals of a former earth.

THE object of enquiry being the operations of the globe, during the time when the present earth was forming at the bottom of the sea, we are now to take a very general view of nature, without descending into those particulars which so often occupy the speculations of naturalists, about the present state of things. We are not at present to enter into any discussion with regard to what are the primary and secondary mountains of the earth; we are not to consider what is the first, and what the last, in those things which now are seen; whatever is most ancient in the strata which we now examine, is supposed to be collecting at the bottom of the sea, during the period concerning which we are now to enquire.

WE have already considered those operations which had been necessary informing our solid land, a body consisting of materials originally deposited at the bottom of the ocean; we are now to investigate the source from whence had come all those materials, from the collection of which the present land is formed; and from knowing the state in which those materials had existed, previously to their entering the composition of our strata, we shall learn something concerning the natural history of this world, while the present earth was forming in the sea.

WE have already observed, that all the strata of the earth are composed either from the calcareous relics of sea animals, or from the collection of such materials as we find upon our shores. at a gross computation, there may perhaps be a fourth part of our solid land, which is composed from the matter that had belonged to those animals. Now, what a multitude of living creatures, what a quantity of animal oeconomy must have been required for producing a body of calcareous matter which is interspersed throughout all the land of the globe, and which certainly forms a very considerable part of that mass! Therefore, in knowing how these animals had lived, or with what they had been fed, we shall have learned a most interesting part of the natural history of this earth; a part which it is necessary to have ascertained, in order to see the former operations of the globe, while preparing the materials of the present land. But, before entering upon this subject, let us examine the other materials of which our land is formed.

GRAVEL forms a part of those materials which compose our solid land; but gravel is no other than a collection of the fragments of solid stones worn round, or having their angular form destroyed by agitation in water, and the attrition upon each other, or upon similar hard bodies. Consequently, in finding masses of gravel in the composition of our land, we must conclude, that there had existed a former land, on which there had been transacted certain operations of wind and water, similar to those which are natural to the globe at present, and by which new gravel is continually prepared, as well as old gravel consumed or diminished by attrition upon our shores.

SAND is the material which enters, perhaps in greatest quantity, the composition of our land. But sand is no other than small fragments of hard and solid bodies, worn or rounded more or less by attrition; consequently, the same natural history of the earth, which is investigated from the masses of gravel, is also applicable to those masses of sand which we find forming so large a portion of our present land throughout all the earth.

CLAY is now to be considered as the last of those materials of which our strata re composed; but, in order to understand the nature of this ingredient, something must be premised.

CLAY is a mixture of different earths or hard substances, in an impalpable state. Those substances are chiefly the siliceous and aluminous earths. Other earths are occasionally mixed in clays, or perhaps always to be found in some small portion. But this does not affect the general character of clay; it only forms a special variety in the subject. A sensible or considerable portion of calcareous earth, in the composition of clay, constitutes a marl, and s sufficient admixture of sand, a loam.

AN indefinite variety of those compositions of clay form a large portion of the present strata, all indurated and consolidated in various degrees; but this great quantity of siliceous, argillaceous, and other compound substances, in form of earth or impalpable sediment, corresponds perfectly with that quantity of those same substances which must have been prepared in the formation of so much gravel and sand, by the attrition of those bodies in the moving waters.

THEREFORE, from the consideration of those materials which compose the present land, we have reason to conclude, that, during the time this land was forming, by the collection of its materials at the bottom of the sea, there had been a former land containing materials similar to those which we find at present in examining the earth. We may also conclude, that there had been operations similar to those which we now find natural to the globe, and necessarily exerted in the actual formation of gravel, sand and clay. But what we have now chiefly to view to illustrate is this, that there had then been in the ocean a system of animated beings, which propagated their species, and which have thus continued their several races to this day.

IN order to be convinced of that truth, we have but to examine the strata of our earth, in which we find the remains of animals. In this examination, we not only discover every genus of animal which at present exists in the sea, but probably every species, and perhaps some species with which at present we are not acquainted. There are, indeed, varieties in those species, compared with the present animals which we examine, but no greater varieties than may perhaps be found among the same species in the different quarters of the globe. Therefore, the system of animal life, which had been maintained in the ancient sea, had not been different from that which now subsists, and of which it belongs to naturalists to know the history. IT is the nature of animal life to be ultimately supported from matter of vegetable production. Inflammable matter may be considered as the *pabulum* of life. This is prepared in the bodies of living plants, particularly in their leaves exposed to the sun and light. This inflammable matter, on the contrary, is consumed in animal bodies, where it produces heat or light, or both. Therefore, however animal matter, or the pabulum of life, may circulate through a series of digesting powers, it is constantly impaired or diminishing in the course of this oeconomy, and, without the productive power of plants, it would finally be extinguished.

THE animals of the former world must have been sustained during indefinite successions of ages. The mean quantity of animal matter, therefore, must have been preserved by vegetable production, and the natural waste of inflammable substance repaired with continual addition; that is to say, the quantity of inflammable matter necessary to the animal consumption, must have been provided by means of vegetation. Hence we must conclude, that there had been a world of plants, as well an ocean replenished with living animals.

WE are now, in reasoning from principles, come to a point decisive of the question, and which will either confirm the theory, if it be just, or confute our reasoning, if we have erred. Let us, therefore, open the book of Nature, and read in her records, if there had been a world bearing plants, at the time when this present world was forming at the bottom of the sea.

HERE the cabinets of the curious are to be examined; but here some caution is required, in order to distinguish things perfectly different, which sometimes are confounded.

FOSSIL WOOD, to naturalists in general, is wood dug up from under ground, without enquiring whether this had been the production of the present earth, or that which had preceded it in the circulation of land and water. The question is important, and the solution of it is, in general, easy. The vegetable productions of the present earth, however deep they may be found buried beneath the surface, and however ancient they may appear, compared with the records of our known times, are new, compared with the solid land on which they grew; and they are only covered with the produce of a vegetable soil, or the alluvion of the present land on which we dwell, and on which they had grown. But the fossil bodes which from the present subject of enquiry, belonged to former land, and are found only in the sea-born strata of our present earth. It is to these alone that we appeal, in order to prove the certainty of former events.

MINERALIZED wood, therefore, is the object now enquired after; that wood which had been lodged in the bottom of the sea, and there composed part of a stratum, which hitherto we have considered as only formed of the materials proper to the ocean. Now, what a profusion of this species of fossil wood is to be found in the cabinets of collectors, and even in the hands of lapidaries, and such artificers of polished stones! In some places, it would seem to be as common as the agate.

I SHALL only mention a specimen in my own collection. It is wood petrified with calcareous earth, and mineralized with pyrites. This specimen of wood contains in itself, even without the stratum of stone in which it is embedded, the most perfect record of its genealogy. It had been eaten or perforated by those sea-worms which destroy the bottoms of our ships. There is the clearest evidence of this truth. Therefore, this wood had grown upon land which stood above the level of the sea, while the present land was only forming at the bottom of the ocean.

WOOD is the most substantial part of plants, as shells are the more permanent part of marine animals. It is not, however, the woody part alone of the ancient vegetable world that is transmitted to us in the record of our mineral pages. We have the type of many species of foliage, and even of the most delicate flower; for, in this way, naturalists have determined, according to the Linnaean system, the species, or at least the genus, of the plant. Thus, the existence of a vegetable system at the period now in contemplation, so far from being doubtful, is a matter of physical demonstration.

THE profusion of this vegetable matter, delivered into the ocean, which then generated land, is also evidenced in the amazing quantities of mineral coal, which is to be found in perhaps every region of the earth.

NOTHING can be more certain, than that all the coaly or bituminous strata have had their origin from the substance of vegetable bodies that grew upon the land. Those strata, though, in general, perfectly consolidated, often separate horizontally in certain places; and there we find the fibrous or vascular structure of the vegetable bodies. Consequently, there is no doubt of fossil coal being a substance of vegetable production, however animal substances also may have contributed in forming this collection of oleaginous or inflammable matter.

HAVING thus ascertained the state of a former earth, in which plants and animals had lived, as well as the gradual production of the present earth, composed from the materials of a former world, it must be evident, that here are two operations which are necessarily consecutive. The formation of the present earth necessarily involves the destruction of the continents in the ancient world; and, by pursuing in our minds the natural operations of a former earth, we clearly see the origin of that land, by the fertility of which, we, and all the animated bodies of the sea, are fed. It is in like manner, that, contemplating the present operations of the globe, we may perceive the actual existence of those productive causes, which are now laying the foundation of land in the unfathomable regions of the sea, and which will, in time, give birth to future continents.

BUT though, in generalizing the operations of nature, we have arrived at those great events, which, at first sight, may fill the mind with wonder and with doubt, we are not to suppose, that there is any violent exertion of power, such as is required in order to produce a great event in little time; in nature, we find no deficiency in respect of time, nor any limitation with regard to power. But time is not made to flow in vain; nor does there ever appear the exertion of superfluous power, or the manifestation of design, not calculated in wisdom to effect some general end.

THE events now under consideration may be examined with a view to see this truth; for it may be enquired, why destroy one continent in order to erect another? The answer is plain; Nature does not destroy a continent from having wearied of a subject which had given pleasure, or changed her purpose, whether for a better or a worse; neither does she erect a continent of land among the clouds, to shew her power, or to amaze the vulgar man: Nature has contrived to productions of vegetable bodies, and the sustenance of animal life, to depend upon the gradual but sure destruction of a continent; that is to say, these two operations necessarily go hand in hand. But with such wisdom has nature ordered things in the oeconomy of this world, that the destruction of one continent is not brought about without the renovation of the earth in the production of another; and the animal and vegetable bodies, for which the world above the surface of the sea is levelled with its bottom, are among the means employed in those operations, as well as the sustenance of those living beings is the proper end in view.

THUS, in understanding the proper constitution of the present earth, we are led to know the source from whence had come all the materials which nature had employed in the construction of the world which appears; a world contrived in consummate wisdom for the growth and habitation of a great diversity of plants and animals; and a world peculiarly adapted to the purposes of man, who inhabits all its climates, who measures its extent, and determines its productions at his pleasure.

THE whole of a great object or event fills us with wonder and astonishment, when all the particulars, in the succession of which the whole had been produced, may be considered without the least emotion. When, for example, we behold the pyramids of Egypt, our mind is agitated with a crowd of ideas that highly entertains the person who understands the subject; but the carrying a heavy stone up to the top of a hill or mountain would give that person little pleasure or concern. We wonder at the whole operation of the pyramid, but not at any one particular part.

THE raising up of a continent of land from the bottom of the sea, is an idea that is too great to be conceived easily in all the parts of its operation, many of which are perhaps unknown to us; and without being properly understood, so great an idea may appear like a thing that is imaginary. In like manner, the co-relative, or corresponding operation, the destruction of the land, is an idea that does not easily enter into the mind of man in its totality, although he is daily witness to part of the operation. We never see a river in a flood, but we must acknowledge the carrying away of part of our land, to be sunk at the bottom of the sea; we never see a storm upon the coast, but we are informed of a hostile attack of the sea upon our country; attacks which must, in time, wear away the bulwarks of our soil, and sap the foundations of our dwellings. Thus, great things are not understood without the analyzing of many operations, and the combination of time with many events happening in succession.

LET us now consider what is to be the subject of examination, and where it is that we are to observe those operations which must determine either the stability or the instability of this land on which we live. OUR land has two extremities; the tops of mountains, on the one hand, and the sea-shores, on the other: it is the intermediate space between these two, that forms the habitation of plants and animals. While there is a sea-shore and a higher ground, there is that which is required in the system of the world: Take these away, and there would remain an aqueous globe, in which the world would perish. But, in the natural operations of the world, the land is perishing continually; and this is that which now we want to understand.

UPON the one extremity of our land, there is no increase, or there is no accession of any mineral substance. That place is the mountain top, on which nothing is observed but continual decay. The fragments of the mountain are removed in a gradual succession from the highest station to the lowest. Being arrived at the shore, and having entered the dominion of the waves, in which they find perpetual agitation, these hard fragments, which had eluded the resolving powers natural to the surface of the earth, are incapable of resisting the powers here employed for the destruction of the land. By the attrition of one hard body upon another, the moving stones and rocky shore, are mutually impaired. And that solid mass, which of itself had potential stability against the violence of the waves, affords the instruments of its own destruction, and thus gives occasion to its actual instability.

IN order to understand the system of the heavens, it is necessary to connect together periods of measured time, and the distinguished places of revolving bodies. It is thus that system may be observed, or wisdom, in the proper adapting of powers to an intention. In like manner, we cannot understand the system of the globe, without seeing that progress of things which is brought about in time, thus measuring the natural operations of the earth with those of the heavens. This is properly the business of the present undertaking. OUR object is to know the time which had elapsed since the foundation of the present continent had been laid at the bottom of the ocean, to the present moment in which we speculate on these operations. The space is long; the data for the calculations are, perhaps, deficient: no matter; so far as we know our error, or the deficiency in our operation, we proceed in science, and shall conclude in reason. It is not given to man to know what things are truly in themselves, but only what those things are in his thought. We seek not to know the precise measure of any thing; we only understand the limits of a thing, in knowing what it is not, either on the one side or the other.

WE are investigating the age of the present earth, from the beginning of that body which was in the bottom of the sea, to the perfection of its nature, which we consider as in the moment of our existence; and we have necessarily another area, which is collateral, or correspondent, in the progress of those natural events. This is the time required, in the natural operations of this globe, for the destruction of a former earth; and earth equally perfect with the present, and an earth equally productive of growing plants and living animals. Now, it must appear, that, if we had a measure for the one of those corresponding operations, we would have an equal knowledge of the other.

THE formation of a future earth being in the bottom of the ocean, at depths unfathomable to man, and in regions far beyond the reach of his observation, here is apart of the process which cannot be taken as a principle in forming an estimate of the whole. But, in the destruction of the present earth, we have a process that is performed within the limits of our observation; therefore, in knowing the measure of this operation, we shall find the means of calculating what had passed on a former occasion, as well as what will happen in the composition of a future earth. Let us, therefore, now attempt to make this estimate of time and labour.

THE highest mountain may be levelled with the plain from when it springs, without the loss of real territory in the land; but when the ocean makes encroachment on the basis of our earth, the mountain, unsupported, tumbles with its weight; and with the accession of hard bodies, moveable with the agitation of the waves, gives to the sea the power of undermining farther and farther into the solid basis of our land. This is the operation which is to be measured; this is the mean proportional by which we are to estimate the age of worlds that have terminated, and the duration of those that are but beginning.

BUT how shall we measure the decrease of our land? Every revolution of the globe wears away some part of some rock upon some coast; but the quantity of that decrease, in that measured time, is not a measurable thing. Instead of a revolution of the globe, let us take an age. The age of man does no more in this estimate than a single year. He sees, that the natural course of things is to wear away the coast, with the attrition of the sand and stones upon the shore; but he cannot find a measure for this quantity which shall correspond to time, in order to form an estimate of the rate of this decrease.

BUT man is not confined to what he sees; he has the experience of former men. Let us then go to the Romans and the Greeks in search of a measure of our coasts, which we may compare with the present state of things. Here, again, we are disappointed; their descriptions of the shores of Greece and of Italy, and their works upon the coast, either give no measure of a decrease, or are not accurate enough for such a purpose.

IT is in vain to attempt to measure a quantity which escapes our notice, and which history cannot ascertain; and we might just as well attempt to measure the distance of the stars without a parallax, as to calculate the destruction of the solid land without a measure corresponding to the whole.

THE description which POLYBIUS has given of the Pontus Euxinus, with the two opposite Bosphori, the Meotis, the Propontis, and the Port of Byzantium, are as applicable to the present state of things, as they were at the writing of that history. The filling up of the bed of the Meotis, an event which, to POLYBIUS, appeared not far off, must also be considered as removed to a very distant period, though the causes still continue to operate as before.

BUT there is a thing in which history and the present sate of things do not agree. It is upon the coast of Spain, where POLYBIUS says there was an island in the mouth of the harbour of New Carthage. At present, in place of the island, there is only a rock under the surface of the water. It must be evident, however, that the loss of this small island affords no proper ground of calculation for the measure or rate of wasting which could correspond to the coast in general; as neither the quantity of what is now lost had been measured, nor its quality ascertained.

LET us examine places much more exposed to the fury of the waves and currents than the coast of Carthagena, the narrow fretum, for example, between Italy and Sicily. It does not appear, that this passage is sensibly wider than when the Romans first had known it. The Isthmus of Corinth is also apparently the same at present as it had been two or three thousand years ago. Scilla and Charibdis remain now, as they had been in ancient times, rocks hazardous for coasting vessels which had to pass that strait.

IT is not meant by this to say, these rocks have not been wasted by the sea, and worn by the attrition of moving bodies, during that space of time; were this true, and that those rocks, the bulwarks of the land upon those coasts, had not been at all impaired from that period, they might remain for ever, and thus the system of interchanging the place of sea and land upon this globe might be frustrated. It is only meant to affirm, that the quantity which those rocks, or that coast, have diminished from the period of our history, has either been too small a thing for human observation, or, which is more probable, that no accurate measurement of the subject, by which this quantity of decrease might have been ascertained, had been taken and recorded. It must be also evident, that a very small operation of an earthquake would be sufficient to render every means of information, in this manner of mensuration, unsatisfactory or precarious.

PLINY says Italy was distant from Sicily a mile and a half; but we cannot suppose that this measure was taken any otherwise than by computation, and such a measure is but little calculated to afford us the just means of a comparison with the present distance. He also says, indeed, that Sicily had been once joined with Italy. His words are: " Quondam BRUTIO agro cohaerens, mox interfuso mari avulsa[9]." But all that we can conclude from this history of PLINY is, that, in all times, to people considering the appearances of those two approached coasts, it had seemed probably, that the sea formed a passage between the two countries which had been once united; in like manner as is still more immediately perceived, in that smaller disjunction which is made between the island of Anglesey and the continent of Wales.

THE port of Syracuse, with the island which forms the greater and lesser, and the fountain of Arethusa, the water of which the ancients divided from the sea with a wall, do not seem to be altered. From Sicily

---

[9] * Lib. 3. cap.8.

to the coast of Egypt, there is an uninterrupted course of sea for a thousand miles; consequently, the wind, in such a stretch of sea, should bring powerful waves against those coasts. But, on the coast of Egypt, we find the rock on which was formerly built the famous tower of Pharos; and also, at the eastern extremity of the port Eunoste, the sea-bath, cut in the solid rock upon the shore. Both those rocks, buffeted immediately with the waves of the Mediterranean sea, are, to all appearance, the same at this day as they were in ancient times.[10]

MANY other such proofs will certainly occur, where the different parts of those coasts are examined by people of observation and intelligence. But is enough for our present purpose, that this decrease of the coasts in general has not been observed; and that it is as generally thought, that the land is gaining upon the sea, as that the sea is gaining upon the land.

TO sum up the argument, we are certain, that all the coasts of the present continents are wasted by the sea, and constantly wearing away upon the whole; but this operation is so extremely slow, that we cannot find a measure of the quantity in order to form an estimate. Therefore, the present continents of the earth, which we consider as in a state of perfection, would, in the natural operations of the globe, require a time indefinite for their destruction.

BUT, in order to produce the present continents, the destruction of a former vegetable world was necessary; consequently, the production of our present continents must have required a time which is indefinite. In like manner, if the former continents were of the same nature as the present, it must have required another space of time, which also is indefinite, before they had come to their perfection as a vegetable world.

WE have been representing the system of this earth as proceeding with a certain regularity, which is not perhaps in nature, but which is necessary for our clear conception of the system of nature. The system of nature is certainly in rule, although we may now know every circumstance of its regulation. We are under a necessity, therefore, of making regular suppositions, in order to come at certain conclusions which may be compared with the present state of things.

IT is not necessary that the present land should be worn away and wasted, exactly in proportion as new land shall appear; or, conversely, that an equal proportion of new land should always be produced as the old is made to disappear. It is only required, that, at all times, there should be a just proportion of land and water upon the surface of the globe, for the purpose of a habitable world.

NEITHER is it required in the actual system of this earth, that every part of the land should be dissolved in its structure, and worn away by

---

[10] Lettres sur l'Egypte, M. SAVARY.

attrition, so as to be floated in the sea. Parts of the land may often sink in a body below the level of the sea, and parts again may be restored, without waiting for the general circulation of land and water, which proceeds with all the certainty of nature, but which advances with an imperceptible progression. Many of such apparent irregularities may appear, without the least infringement on the general system. That system is comprehended in the preparation of future land at the bottom of the ocean, from those materials which the dissolution and attrition of the present land may have provided, and from those which the natural operations of the sea afford.

IN thus accomplishing a certain end, we are not to limit nature with the uniformity of an equable progression, although it be necessary in our computations to proceed upon equalities. Thus also, in the use of means, we are not to prescribe to nature those alone which we think suitable for the purpose, in our narrow view. It is our business to learn of nature (that is by observation) the ways and means, which in her wisdom are adopted; and we are to imagine these only in order to find means for further information, and to increase our knowledge from the examination of things which actually have been. It is in this manner, that intention may be found in nature; but this intention is not to be supposed, or vainly imagined, from what we may conceive to be.

WE have been now supposing that the beginning of our present earth had been laid in the bottom of the ocean, at the completion of the former land; but this was only for the sake of distinctness. The just view is this, that when the former land of the globe had been complete, so as to begin to waste and be impaired by the encroachment of the sea, the present land began to appear above the surface of the ocean. In this manner we suppose a due proportion to be always preserved of land and water upon the surface of the globe, for the purpose of a habitable world, such as this which we possess. We thus, also, allow time an opportunity for the translation of animals and plants to occupy the earth.

BUT, if the earth on which we live, began to appear in the ocean at the time when the last began to be resolved, it could not be from the materials of the continent immediately preceding this which we examine, that the present earth had been constructed ; for the bottom of the ocean must have been filled with materials before land could be made to appear above its surface.

LET us suppose that the continent, which is to succeed our land, is at present beginning to appear above the water in the middle of the Pacific Ocean, it must be evident, that the materials of this great body, which is formed and ready to be brought forth, must have been collected from the destruction of an earth which does not now appear. Consequently, in this true statement of the case, there is necessarily required the destruction of an animal and vegetable earth prior to the former land; and the materials of that earth which is first in our account, must have been collected at the bottom of the ocean, and begun to be concocted for the production of the present earth, when the land immediately preceding the present had arrived at its full extent. THIS, however, alters nothing with regard to the nature of those operations of the globe. The system is still the same. It only protracts the indefinite space of time in its existence, while it gives us a view of another distinct period of the living world; that is to say, the world which we inhabit is composed of the materials, not of the earth which was the immediate predecessor of the present, but of the earth which, in ascending from the present, we consider as the third, and which had preceded the land that was above the surface of the sea, while our present land was yet beneath the water of the ocean. Here are three distinct successive periods of existence, and each of these is, in our measurement of time, a thing of indefinite duration.

WE have now got to the end of our reasoning; we have no data further to conclude immediately from that which actually is: But we have got enough; we have the satisfaction to find, that in nature there is wisdom, system, and consistency. For having, in the natural history of this earth, seen a succession of worlds, we may from this conclude that there is a system in nature; in like manner as, from seeing revolutions of the planets, it is concluded, that there is a system by which they are intended to continue those revolutions. But if the succession of worlds is established in the system of nature, it is in vain to look for any thing higher in the origin of the earth. The result, therefore, of our present enquiry is, that we find no vestige of a beginning,--no prospect of an end.

Printed in Great Britain
by Amazon

63568774R00047